O Level Chemistry for Cambridge Students: 2024 and 2025 Edition

Azhar ul Haque Sario

Copyright

First Printing, 2024

Azhar.sario@hotmail.co.uk

ORCID: https://orcid.org/0009-0004-8629-830X

Disclaimer: This book is free from AI use. The cover was designed in Microsoft Publisher. This book covers the complete syllabus for the Cambridge O Level Chemistry 2024-to-2025-year syllabus. It is the author's original work and has not been copied from other sources. It is intended as a supplementary resource for exam preparation.

Contents

States of matter

Solids, liquids and gases

Imagine a world where:

Solids are like stubborn old men, refusing to change their ways or budge an inch. They're set in their routines, with their molecules holding hands like a tightly-knit community.
Liquids are like playful children, always moving and adapting to their surroundings. They love to explore and take the shape of whatever container they find themselves in, like a chameleon changing its colors.
Gases are like free spirits, wild and untamed. They have no boundaries and love to spread out and fill any space they can find, like a group of teenagers at a party.

Let's dive deeper into this whimsical world of matter:

Solids:

Picture a diamond: Its atoms are locked in a rigid embrace, creating a structure so strong it can cut through glass. This is why solids hold their shape and are tough to compress.
Think of a steel beam: Its molecules are like disciplined soldiers, standing in formation and resisting any attempt to bend them out of shape. This is why solids are used in construction to provide stability and support.

Liquids:

Imagine a river: Its water molecules flow and adapt to the curves and bends of the riverbed, constantly changing shape yet maintaining a consistent volume. This is why liquids can be poured and take the shape of their container.
Think of honey: Its molecules are like dancers, gracefully moving around each other but still maintaining a close connection. This is why

liquids have a moderate density and can flow, but not as freely as gases.

Gases:

Picture the air we breathe: Its molecules are like a swarm of bees, buzzing around randomly and filling every nook and cranny. This is why gases expand to fill their container and have no fixed shape or volume.
Think of a hot air balloon: The heat causes the air molecules inside to become even more energetic and spread out, making the balloon rise. This is why gases are highly compressible and have low density.

Let's explore some fascinating case studies:

Water: This magical substance can transform between all three states of matter, like a master of disguise. As ice, its molecules are in a fixed formation, creating a crystalline structure. As liquid water, they loosen up and dance around, allowing it to flow. As steam, they become wild and free, expanding to fill the space.
Carbon dioxide: This gas can be captured and transformed into a solid called dry ice, which is used to keep things cold. It's like freezing the air itself! The dry ice then magically disappears as it turns directly back into a gas, like a magician vanishing into thin air.
Liquid crystals: These substances are like chameleons, exhibiting properties of both liquids and solids. They can flow like liquids but also maintain some order in their molecular arrangement, allowing them to be used in LCD screens.

In conclusion:

The world of matter is a fascinating place, full of unique characters and surprising transformations. By understanding the distinct properties and structures of solids, liquids, and gases, we can unlock the secrets of the universe and appreciate the magic that surrounds us every day.

Imagine a bustling dance floor. The dancers (molecules) in their fancy attire (states of matter) are moving to the music (heat energy).

Solid State: The dancers are in a tightly packed formation, swaying gently to a slow waltz. They're holding hands (strong attraction), not wanting to lose their partners.

Melting: The DJ cranks up the tempo! The dancers start to move with more energy, shaking and twisting. Some let go of their partners and start to glide around the floor. The dance floor is now a mix of slow dancers and more energetic movers.

Liquid State: The music shifts to a lively salsa! The dancers are now moving freely, twirling and dipping, bumping into each other but not sticking together for long.

Boiling: The DJ switches to an electrifying techno beat! The dancers are jumping and spinning with wild abandon. They break free from the crowd and leap into the air, soaring like they're in a mosh pit.

Gaseous State: The dancers are now floating in a vast, open space, moving in every direction with incredible speed. They occasionally bump into each other and the walls, but they're mostly free to explore.

Cooling: The music slows down, and the dancers gradually lose their energy. They start to come closer together, drawn by an invisible force (attraction).

Condensation: As the music shifts back to a slow rhythm, the dancers pair up again, holding onto each other tightly. They're still moving, but they're now confined to a smaller space.

Freezing: The music fades to a gentle lullaby. The dancers come to a standstill, forming a tightly packed, organized group. They're still swaying slightly, but they're no longer moving around.

Heating and Cooling Curves: Think of these as a visual representation of the music's intensity throughout the night. The steeper the curve, the faster the music changes. The flat sections represent moments when the DJ changes the song, and the dancers adjust to the new rhythm.

Effect of Temperature and Pressure: Imagine the dance floor shrinking (increased pressure) or expanding (decreased pressure). How would that affect the dancers' movements? What if the room got hotter (increased temperature) or colder (decreased temperature)?

Haber Process: This is like a choreographed dance where nitrogen and hydrogen molecules are the dancers. The DJ (catalyst) sets the tempo (temperature and pressure) to ensure the dancers move in the right way to create a new formation (ammonia).

Conclusion: The kinetic particle theory is like a universal dance language that explains how matter behaves in different situations. By understanding this language, we can unlock the secrets of the universe and create new technologies that benefit humanity.

Diffusion

Imagine a bustling city: crowds rushing, cars honking, a whirlwind of movement. That's kind of like what's happening on a microscopic level when we talk about diffusion. Tiny particles, like mischievous sprites, are zipping around, bumping into each other and bouncing off walls. They're driven by an insatiable urge to spread out, to explore every nook and cranny. This inherent restlessness is what we call kinetic energy, the lifeblood of diffusion.

Think of it like this: You've just sprayed your favorite perfume in a room. At first, the scent is strong where you sprayed it. But slowly, those fragrant molecules are mingling with the air, spreading out like gossip in a schoolyard. Eventually, the whole room is filled with the aroma. That's diffusion in action!

Now, these tiny travelers aren't all created equal. Some are like nimble ballerinas, flitting about with grace. Others are more like lumbering giants, slow and steady. This is where molecular mass comes into play. Imagine a race between a feather and a bowling ball. The feather, light and airy, will dance on the breeze, while the bowling ball, heavy and grounded, will lag behind. It's the same with molecules: the lighter they are, the faster they can zip around and spread out.

This principle is captured in something called Graham's Law, a sort of molecular speed limit. It tells us that lighter molecules are the speed demons of the diffusion world. They're like the cheetahs of the microscopic savanna, while the heavier molecules are more like the elephants, powerful but ponderous.

But why does this matter? Well, diffusion is happening all around us, and even inside us! It's how our lungs take in oxygen and release carbon dioxide. It's how nutrients spread through our bodies and how pollutants disperse in the environment. Understanding how molecular mass affects diffusion helps us understand these vital processes and even harness them for technological advancements.

For instance, scientists used this principle to separate different types of uranium during the Manhattan Project, a pivotal moment in history. Today, researchers are exploring how diffusion can be used to deliver drugs more effectively, create new materials with amazing properties, and even clean up environmental contamination.

So next time you see a whiff of smoke drifting through the air or smell the aroma of freshly baked bread, remember the invisible dance of molecules, driven by their restless energy and governed by their mass. It's a tiny world with a big impact, a constant reminder that even the smallest things can make a world of difference.

Atoms, elements and compounds

Elements, compounds and mixtures

Imagine the universe as a giant LEGO set.

Elements are like the individual LEGO bricks. They're the simplest building blocks, each with its own unique color, shape, and size (properties). You've got your classic red bricks (hydrogen), sturdy blue ones (oxygen), and maybe some sparkly gold ones (guess what!). You can't break these bricks down any further – they're the pure essentials.

Compounds are like the amazing things you build with those LEGOs. Combine a few red bricks and a blue one in just the right way, and voilà! You've got water (H2O). Or maybe you stack some black bricks (carbon) in a special pattern and create a diamond. The key is that compounds are brand new things with their own unique properties, different from the individual bricks they're made of.

Mixtures are like a big box of LEGOs, all jumbled together. You've got all sorts of bricks in there, but they're not connected in any specific way. You can easily pick them out and sort them. That's like a mixture – you can separate the different parts (like sand and water) because they haven't chemically bonded together.

Think of it like baking a cake:

Elements are your ingredients: flour, sugar, eggs, etc.
Compounds are the cake batter: You've mixed the ingredients together, and they've chemically reacted to create something new.
Mixtures are the toppings: Sprinkles, chocolate chips, and frosting are all mixed together on top, but you can easily pick them apart.
Why should you care about this stuff?

Well, understanding elements, compounds, and mixtures is like having the key to unlock the secrets of the universe! It helps us understand everything from why the sky is blue to how our bodies work. It's the foundation of chemistry, and it's essential for solving some of the world's biggest challenges, like developing new medicines and creating sustainable energy sources.

So, next time you look around, remember that everything you see is made up of these tiny building blocks. Pretty cool, huh?

Atomic structure and the Periodic Table

Journey to the Heart of Matter: Unraveling the Atom's Secrets

Imagine a world built on tiny, invisible LEGO bricks. That's essentially what atoms are – the fundamental building blocks of everything around us. But unlike LEGOs, atoms are far more intricate and mysterious, holding the key to understanding how the universe works at its most basic level.

Delving into the Atomic Core

At the center of every atom lies a bustling metropolis – the nucleus. This incredibly dense core is like a tiny, tightly packed solar system, containing nearly all of the atom's mass. Within this bustling hub, we find two types of particles:

Protons: The positively charged citizens, like tiny suns radiating energy. The number of protons determines the atom's identity, like a cosmic ID card.
Neutrons: The neutral peacekeepers, adding to the mass without affecting the atom's personality.
These particles are held together by an incredibly strong force, like an invisible glue that keeps the nucleus from flying apart.

The Ethereal Electron Cloud

Surrounding the nucleus is a hazy cloud of negatively charged electrons, like a swarm of bees buzzing around a hive. These electrons don't follow fixed orbits like planets; instead, they exist in a quantum dance, their positions and movements governed by the laws of probability.

Think of it like a concert hall, with electrons occupying different energy levels, like rows of seats. These levels are further divided into sections, or subshells, each with its own unique energy signature.

Quantum Mechanics: The Maestro of the Atom

To truly understand the atom, we need to enter the realm of quantum mechanics – a mind-bending world where particles can behave like waves and uncertainty reigns supreme. This is where the true magic of the atom unfolds, revealing its secrets through:

Quantum Numbers: These are like the electron's address and job description, specifying its energy level, shape, and orientation.
Electron Configuration: This is the atom's seating chart, showing how electrons are arranged in their quantum "seats."
Hund's Rule and the Pauli Exclusion Principle: These are the rules of the quantum concert hall, ensuring that electrons occupy their seats in an orderly and predictable fashion.

Atomic Spectra: The Atom's Fingerprint

When atoms get excited, they release energy in the form of light, like a celestial fireworks display. Each element has its own unique light signature, or atomic spectrum, like a fingerprint that reveals its identity.

Case Studies: Peeking into the Atomic World

The Hydrogen Atom: The simplest of all atoms, with just one proton and one electron, hydrogen is the Rosetta Stone of quantum mechanics, helping us decipher the atom's code.
The Carbon Atom: The backbone of life itself, carbon's ability to form four bonds makes it the ultimate molecular architect, creating the complex molecules that make up living organisms.
The Sodium Atom: This eager electron donor plays a crucial role in biological processes, like transmitting nerve impulses.

The Atom's Legacy: From Medical Marvels to Cosmic Mysteries

Understanding the atom has led to incredible advancements in medicine, technology, and our understanding of the universe. Medical imaging techniques, like X-rays and MRI, rely on the interaction of radiation with atoms in our bodies. And by studying the atomic spectra of distant stars, we can unravel the mysteries of their composition and evolution.

The Journey Continues

The atom is a microcosm of the universe, a world of endless wonder and complexity. As we continue to explore its depths, we unlock new secrets and gain a deeper appreciation for the intricate machinery that drives the cosmos.

Imagine the atom as a bustling city:

The Proton Number (Atomic Number) is like the city's unique ID number. It's how you know you're in Atom Ville, Carbon City, or Oxygen polis! This ID number is determined by the protons, which are like the city's founders – the original inhabitants who established its identity. They live right in the heart of the city, in the nucleus – the city center where all the important decisions are made.

In a peaceful, balanced city, the number of protons (founders) equals the number of electrons (citizens). The citizens are always on the move, buzzing around the city center (nucleus) like they're on busy highways. As long as the number of founders and citizens is equal, the city remains neutral, with no overall charge.

But sometimes, citizens move in or out of the city, and things get a little chaotic! If citizens leave, the city becomes positively charged (a cation) – think of it as having more "pep" or energy. If citizens move in, the city becomes negatively charged (an anion) – maybe it's a bit more relaxed and "chill" now.

Even within the same city, you can have different neighborhoods (isotopes). These neighborhoods have the same number of founders (protons) and the same city ID (atomic number), but they have different numbers of neutrons, which are like the city's buildings. Some neighborhoods have more buildings, some have fewer, making them slightly different in character, even though they're part of the same city.

Now, imagine a grand map of all these atomic cities: the Periodic Table! Cities are grouped together based on their similarities – like cities with a love for the arts, or cities known for their industry. These groups (columns) share common traits because they have the same number of citizens living in their outer suburbs (valence electrons).

Think of it like this:

Hydrogen (H): A tiny village with just one founder (1 proton).
Helium (He): A slightly larger town with two founders (2 protons).
Oxygen (O): A bustling metropolis with eight founders (8 protons).
Gold (Au): A grand, ancient city with a rich history, boasting 79 founders (79 protons)!

Isotopes in Medicine:

Imagine a special team of doctors who use isotopes like tiny, targeted missiles to fight diseases! For example, iodine-131 is like a tiny warrior that goes straight to the troublemaker in the thyroid gland and delivers a powerful blow to defeat the enemy (cancer cells).

Mass Number:

If the proton number is like a city's ID, the mass number is like knowing the city's total population – founders (protons) plus buildings (neutrons). It gives you a good idea of the city's overall size and weight.

Carbon Dating:

Think of carbon-14 as a tiny clock within ancient artifacts. Over time, this clock "ticks" slower and slower as the carbon-14 decays. By measuring how much the clock has slowed down, scientists can figure out how much time has passed since the artifact was alive.

In Conclusion:

The proton number and mass number are like the essential characteristics that define each atomic city, giving them their unique identity and properties. They're the keys to understanding the amazing diversity and complexity of the atomic world!

Let's dive into the fascinating world of atoms and how their electrons are arranged! Imagine each atom as a bustling city, and the electrons are the residents living in different neighborhoods (shells) and apartments (subshells).

1. Electron City: Population 1 to 20

Think of the nucleus, the atom's center, as the downtown area. The first shell, closest to downtown, has only enough room for two electron residents. The second and third shells, a bit further out, can each house up to 8 electrons.

So, for the first 20 elements on the periodic table, we fill up these electron cities like this: 2, 8, 8, 2. It's like a building code for atoms!

Here's a peek into some of these electron cities:

Hydrogen (H): A tiny town with just 1 electron.
Helium (He): A cozy village with 2 electrons, filling up the first shell.
Lithium (Li): A growing town with 3 electrons, 2 in the first shell and 1 starting to fill the second.
Neon (Ne): A bustling city with all its shells full – 2, 8!
Potassium (K): A sprawling metropolis with 19 electrons, even starting to populate a fourth shell!

2. The Periodic Table: A Map of Electron Cities

The periodic table is like a map of all these electron cities, organized in a clever way.

Group VIII: The Exclusive Neighborhood

Group VIII is like an exclusive, gated community. The elements here, the noble gases (like Helium, Neon, and Argon), have their outer shells completely full. They're content and don't like to interact much with other elements – hence the name "noble." 👑

Groups I to VII: The Social Butterflies

Elements in Groups I to VII are more social. Their outer shell residents (valence electrons) are always looking to mingle and form bonds with electrons from other atoms. The group number tells you exactly how many "social butterfly" electrons each element has. 🦋

Periods: Circles on the Map

The periods (rows) on the periodic table are like concentric circles on our map, each representing a new shell being filled. Hydrogen and Helium are in the first circle, Lithium to Neon in the second, and so on. ⭕

3. Ions: Electron Movers and Shakers

Sometimes, electron residents decide to move! If an atom loses an electron, it becomes a positive ion (cation) – like a city losing a resident. If it gains an electron, it becomes a negative ion (anion) – like a city welcoming a newcomer. 🚚

4. Why Does This Matter?

Understanding how electrons are arranged helps us explain a lot about how elements behave:

Chemical Bonding: How atoms form relationships with each other (like electron handshakes or sharing apartments).

Reactivity: How eager an element is to make friends (or react) with others.
Physical Properties: Things like melting point, boiling point, and whether an element is a good conductor of electricity.
Periodic Trends: Patterns of behavior across the periodic table – like trends in fashion across different cities!

5. Real-World Examples

Salt (NaCl): Sodium (Na) gives up an electron to chlorine (Cl), like a friendly exchange between neighbors, creating an ionic bond.
Water (H2O): Oxygen (O) shares electrons with two hydrogens (H), like roommates sharing an apartment, forming covalent bonds.

6. In Conclusion

The electronic configuration of an element is like its DNA, revealing its personality and how it interacts with the world. By understanding these electron arrangements, we unlock the secrets of chemistry and the elements!

Isotopes

Isotopes: The Quirky Siblings of the Atomic World

Imagine a family reunion. You've got your aunts and uncles, all a bit different, but sharing that same family resemblance. Now picture atoms like that – they're all part of the same element family (like the 'Carbon' family), but they come in slightly different flavors. Those are isotopes!

Think of protons as the family name. Every carbon atom, no matter what, has six protons – it's their ID card. But then there are the neutrons, the mischievous siblings who love to shake things up. Some carbon atoms have six neutrons, some have seven, some even have eight! That's what makes them isotopes – same family name, different number of those playful neutrons.

Now, you might think, "More neutrons, must mean a totally different personality, right?" Not really! See, the electrons, those tiny whirlwinds orbiting the atom, are the real social butterflies. They determine how an atom interacts with others, and since all isotopes of an element have the same number of electrons, they mostly act the same in the chemical world.

But hold on, there's a twist!

Some isotopes are a bit... unstable. Like that cousin who always seems to be causing a commotion. These are the radioactive isotopes. They have a little too much energy and shed it by emitting tiny particles or bursts of energy. Scientists use these "radioisotopes" in all sorts of cool ways:

Medical Detectives: Radioisotopes are like tiny spies, helping doctors see inside the body with amazing detail. They can track down diseases and even deliver targeted treatments to zap those nasty cancer cells.
Time Travelers: Ever wondered how we know how old dinosaur bones are? Radioactive isotopes act like atomic clocks, ticking away at a steady rate. Scientists can measure this "ticking" to figure out the age of ancient artifacts and even the Earth itself!

Environmental Guardians: Isotopes can help us track pollution, understand climate change, and even clean up contaminated areas. They're like tiny environmental superheroes!

Isotopes in Action: A Tale of Two Cities

Let's take a trip down history lane. The Manhattan Project, a top-secret mission during World War II, harnessed the power of isotopes to create the atomic bomb. It's a story of incredible scientific achievement, but also a stark reminder of the immense responsibility that comes with wielding such power.

On a different note, the Chernobyl disaster showed us the devastating consequences when things go wrong with nuclear technology. It's a cautionary tale, urging us to handle these powerful forces with the utmost care and respect.

The Future of Isotopes: A World of Possibilities

Today, scientists are exploring even more amazing uses for isotopes. Imagine:

Super-precise cancer treatments that deliver radiation directly to tumor cells, sparing healthy tissue.
New materials with incredible properties, like super-strong coatings and self-healing plastics.
Cleaner energy sources and innovative ways to combat climate change.
The world of isotopes is full of surprises. These quirky atomic siblings, with their subtle differences and incredible powers, are shaping our understanding of the universe and driving innovation in countless fields. So, the next time you hear about isotopes, remember – they're not just boring atoms, they're the key to unlocking a world of possibilities!

Journey to the Center of Matter: Atoms, Isotopes, and Ions

Imagine yourself shrinking down, down, down, until you're smaller than a speck of dust. You find yourself in a world of incredible tiny particles – the building blocks of everything around you. Welcome to the realm of atoms!

Atoms: The Tiny Titans

Atoms are like miniature solar systems. At the center is a dense nucleus, home to two types of particles:

Protons: The "positive vibes" guys, giving the nucleus a plus charge.
Neutrons: The "neutral dudes," just chilling with the protons.
Whizzing around the nucleus are the electrons, like tiny planets orbiting a sun. They carry a negative charge, balancing out the protons' positivity.

Atomic Symbol: The Atom's ID Card

Each element has its own unique atomic symbol, like a personal ID card. It tells us:

Atomic Number: The number of protons in the nucleus (the element's "fingerprint").
Mass Number: The combined number of protons and neutrons (the atom's "weight").
For example, carbon's symbol is C, with an atomic number of 6 and a mass number of 12 (most common form). This tells us a carbon atom has 6 protons and 6 neutrons.

Isotopes: The Atom's Siblings

Sometimes, atoms of the same element have a different number of neutrons. These are called isotopes – like siblings with different personalities.

Think of carbon again. It has a few isotopes, like carbon-12 (6 neutrons), carbon-13 (7 neutrons), and carbon-14 (8 neutrons). They're all still carbon, but with slightly different weights.

Ions: The Atom's Charged Cousins

Atoms can also gain or lose electrons, becoming charged particles called ions.

Cations: Atoms that lose electrons, becoming positively charged (like a party animal who's lost their friends).
Anions: Atoms that gain electrons, becoming negatively charged (like someone who's found a lost wallet).
Ions have their own special symbols too, showing their charge. For example, a chlorine atom that gains an electron becomes a chloride ion (Cl^-).

Calculating Relative Atomic Mass: The Atom's Average Weight

Since elements often have multiple isotopes, we use the relative atomic mass (AR) to represent their average weight. It's like calculating the average age of a group of friends.

To find AR, we use a special formula that takes into account the mass and abundance of each isotope. It's like a weighted average, where the more abundant isotopes have a bigger influence on the final result.

Lead: A Case Study

Lead (Pb) has four naturally occurring isotopes, each with its own mass and abundance. By plugging these values into the AR formula, we can calculate lead's relative atomic mass.

Additional Notes: The Fine Print

Relative atomic mass is usually a decimal, reflecting the mix of isotopes.
It's a key concept in chemistry, used in many calculations.
You can find the relative atomic mass of any element on the periodic table.

The End... or is it?

This is just the beginning of your journey into the amazing world of atoms. Keep exploring, keep asking questions, and keep marveling at the tiny titans that make up everything around us!

Ion and ionic bonds

Imagine this: Atoms are like tiny, bustling cities. At the center is a "downtown" nucleus, where the positive protons (think "upbeat citizens") hang out with the neutral neutrons ("the quiet folks"). Zipping around downtown in express lanes are the negative electrons ("busy commuters"). Usually, a city has just the right balance of upbeat citizens and busy commuters, so things are electrically neutral – no one's mood dominates.

But sometimes, an atom-city wants a change of pace. It might lose some of those busy commuters (electrons), becoming a cation – a positively charged city bursting with optimism! Think of metals like sodium and magnesium as those "go-getter" cities, always ready to shed some commuters and embrace a more positive vibe.

Or, an atom-city might attract more commuters (electrons), becoming an anion – a negatively charged city humming with quiet energy. Non-metals like chlorine and oxygen are like those "cozy" cities, happy to welcome more commuters and settle into a more relaxed state.

Why the change? Well, every atom-city wants to be like the "noble gases" – those super-chill cities with the perfect balance (eight commuters in their outer lane). It's the atomic equivalent of finding inner peace!

Now, let's talk relationships! When a positive atom-city (cation) meets a negative atom-city (anion), sparks fly! They're irresistibly drawn to each other, forming a strong bond and creating a massive, intricate structure called a giant ionic lattice. Imagine a vast, 3D chessboard with alternating positive and negative pieces – that's your lattice!

These lattices are super strong (thanks to those intense attractions), which is why ionic compounds have high melting and boiling points. Think of them as incredibly sturdy buildings that can withstand a lot of heat. They're also hard and brittle – like a beautifully crafted glass sculpture, strong but easily shattered if handled roughly.

Examples? Table salt (sodium chloride) is a classic! Imagine tiny cubes made of alternating sodium and chlorine ions, stacked neatly together. Magnesium oxide? Similar story, but with magnesium and oxygen ions instead.

And here's a cool application: Lithium-ion batteries! These power our phones and laptops, and they rely on the movement of lithium ions (those energetic little commuters) between two electrodes. It's like a constant flow of commuters between two bustling cities, keeping the energy flowing!

So, there you have it! Ions and ionic compounds, explained with a touch of humanization, creativity, and a sprinkle of imagination. Hopefully, this makes these concepts a bit more engaging and memorable.

Forget boring textbooks! Let's unravel the secrets of ionic bonds with a sprinkle of pizzazz and a dash of imagination.

Imagine a bustling city filled with atoms, each with its own unique personality. Some atoms, like the social butterflies of the metal world, are always ready to share their electrons. Others, the introverted non-metals, prefer to keep their electrons close. But when these two worlds collide, sparks fly, and ionic bonds are born!

Metals, the generous givers:

Picture a group of friends at a pizza party. The metals, like the generous friend who always brings extra slices, are eager to share their electrons. They lose an electron (or two, or three!), transforming into positive ions, or cations. Think of them as the life of the party, radiating positive energy.

Non-metals, the eager acceptors:

Now imagine the friend who arrives at the party with an empty stomach, ready to devour those extra slices. Non-metals, with their strong appetite for electrons, gladly accept the electrons offered by the metals. They become negative ions, or anions, content with their newfound electron riches.

The electric tango of attraction:

With their opposite charges, the cations and anions are irresistibly drawn to each other, like dancers in an electrifying tango. This powerful attraction forms the ionic bond, a strong and stable connection that holds the atoms together.

Properties that pop!

Ionic compounds, formed by these electric embraces, have some pretty remarkable qualities:

High melting and boiling points: Imagine trying to separate those tangoing dancers – it would take quite a bit of effort! Similarly, breaking the strong ionic bonds requires a lot of energy, resulting in high melting and boiling points.
Electrical conductivity: When dissolved in water or melted, ionic compounds become excellent conductors of electricity. It's like the dance floor suddenly becomes a superhighway for electrons, allowing them to zip around freely.

Real-world wonders:

Ionic compounds aren't just theoretical concepts; they're all around us!

Table salt (NaCl): The star of your dinner table, formed by the passionate bond between sodium and chlorine.
Seashells (CaCO3): These intricate beauties owe their strength and durability to the ionic bonds in calcium carbonate.
Lithium-ion batteries: Powering our smartphones and laptops, these batteries rely on the movement of lithium ions, the tiny dancers that keep our devices running.

So, there you have it – ionic bonds, a captivating tale of electron exchange, electric attraction, and the amazing properties that emerge from this atomic dance. Who knew chemistry could be so much fun?

Simple molecules and covalent bonds

Covalent Bonding: A Molecular Dance of Shared Electrons

Imagine atoms as tiny dancers, each yearning for a complete and balanced ensemble of electrons in their outer shells. In the world of chemistry, this stability is often achieved through a graceful partnership known as covalent bonding. Like dancers holding hands, atoms share electrons, creating a harmonious connection that binds them together.

The Noble Gases: The Stars of Stability

Noble gases, like helium and neon, are the envy of the atomic world. Their outer shells are already filled with electrons, making them content and unreactive. Other atoms, seeking this same state of bliss, form covalent bonds to mimic the electron configuration of these noble gas "stars."

The Electric Tango: Attraction and Repulsion

Covalent bond formation is a delicate dance between attraction and repulsion. The positively charged nuclei of the atoms are drawn to the negatively charged shared electrons, like dancers drawn to a mesmerizing rhythm. However, the nuclei also repel each other, creating a tension that must be balanced for a stable bond to form.

A Chlorine Waltz: Sharing for Stability

Picture two chlorine atoms, each with seven electrons in their outer shells. Like dancers needing a partner to complete their set, they share one electron each, forming a covalent bond. This shared pair of electrons allows both chlorine atoms to achieve a stable octet, mirroring the electron configuration of the noble gas argon.

Simple Molecular Compounds: A Delicate Balance of Forces

Simple molecular compounds are like intricate dances where molecules, the individual dance pairs, are held together by subtle forces. These intermolecular forces are much weaker than the strong covalent bonds within the molecules, like the gentle touch between dance partners compared to the firm grip they have on each other's hands.

Melting and Boiling Points: A Change of Rhythm

The weak intermolecular forces in simple molecular compounds mean that they have low melting and boiling points. It's like changing partners in a dance – it doesn't take much energy to break the connection and move on. When heated, the molecules gain energy and vibrate more vigorously, like dancers getting excited. At the melting point, they have enough energy to overcome the intermolecular forces and break free from their rigid structure, transitioning from a slow waltz to a lively jig. Further heating leads to the boiling point, where the molecules completely overcome the intermolecular forces and fly apart, like dancers taking a final bow and exiting the stage.

Examples:

Water (H2O): Despite its strong hydrogen bonds (a special type of intermolecular force), water has relatively low melting and boiling points. Imagine water molecules as graceful dancers with a strong connection, yet still able to move fluidly between partners.
Carbon dioxide (CO2): This gas at room temperature has weak intermolecular forces, like fleeting connections between dancers in a fast-paced dance.
Methane (CH4): The main component of natural gas, methane has a very low boiling point, reflecting the weak London dispersion forces between its molecules. Think of methane molecules as independent dancers, briefly interacting before moving on.

Electrical Conductivity: A Matter of Charged Partners

Simple molecular compounds are generally poor conductors of electricity. They lack charged particles, or "charged dance partners," that are free to move and carry an electric current. The electrons in covalent bonds are localized between the bonded atoms, like dancers focused on their partner, unable to transmit the electrical flow across the dance floor.

Examples:

Sugar ($C_{12}H_{22}O_{11}$): Sugar crystals are like a ballroom full of dancers in fixed positions, unable to conduct electricity.
Wax (C_nH_{2n+2}): Wax, a mixture of long-chain hydrocarbons, is also a poor conductor, like a dance floor covered in a thick, insulating layer.

Exceptions to the Rule: The Dance of Delocalized Electrons

While most simple molecular compounds are poor conductors, there are exceptions. Graphite, a form of carbon, has a unique layered structure with delocalized electrons that can move freely within the layers, like dancers weaving through a crowded ballroom. This makes graphite a good conductor of electricity.

Conclusion: The Molecular Dance of Life

Covalent bonding and simple molecular compounds are fundamental to the chemistry of life. They are the basis of the intricate molecules that make up our bodies and the world around us. Understanding these concepts is like appreciating the choreography of a complex dance, revealing the beauty and elegance of the molecular world.

Imagine atoms as tiny dancers, each longing to complete their perfect circle of partners. They crave that feeling of stability, of having just the right number of companions in their swirling electron dance. But some atoms are a bit shy, hesitant to give away their precious electrons completely. That's where the magic of covalent bonds comes in!

Think of it as a delightful compromise, a graceful partnership where atoms share their electrons, intertwining their dance moves to create a beautiful, balanced molecule. It's like holding hands with a friend, each of you contributing to a sense of shared joy and stability.

Let's peek into this microscopic ballroom and witness some of these captivating dances:

Hydrogen (H_2): Two shy hydrogen atoms, each with a single electron, come together. They tentatively reach out, their lone electrons intertwining like a delicate pas de deux. Suddenly, they both feel complete, like they've found their perfect dance partner!

Chlorine (Cl_2): These two, already surrounded by a bustling entourage of electrons, still feel a slight longing. They each extend a hand, sharing one electron to form a single bond. Now, with their octet complete, they waltz with newfound confidence and grace.

Oxygen (O_2): Ah, the passionate oxygen duo! They yearn for a deeper connection, a more intense dance. They intertwine not one, but two pairs of electrons, forming a double bond. Their dance is fiery and strong, a testament to their shared bond.

Nitrogen (N_2): These two are truly inseparable, their bond a whirlwind of shared electrons. Three pairs intertwine, creating a triple bond, the strongest connection of all! Their dance is mesmerizing, a testament to the power of shared electrons.

But the dance floor isn't just for couples! Atoms of different elements join in, creating intricate patterns and breathtaking choreography.

Water (H_2O): Oxygen, with its longing for two more electrons, finds the perfect partners in two hydrogen atoms. They form a graceful trio, their dance fluid and life-giving, just like water itself.

Methane (CH_4): Carbon, the versatile atom, takes center stage, extending its four "hands" to connect with four eager hydrogens. Their dance is energetic and playful, forming the building block of countless organic molecules.

And so, the dance goes on, atoms twirling and bonding, creating the wondrous diversity of molecules that make up our world. From the air we breathe to the intricate machinery of our cells, covalent bonds are the elegant choreography of life itself.

Want to delve deeper into this fascinating world? Let's explore the concepts of electronegativity, bond length, and molecular geometry – the nuances that add even more beauty and complexity to this atomic dance. We'll even uncover the secrets behind the dazzling diamond and the life-giving DNA molecule!

Giant covalent structures

Imagine a world built on tiny, indestructible LEGO bricks. That's essentially what a giant covalent structure is – a vast, intricate network of atoms held together by super-strong bonds. These structures give rise to some of the most remarkable materials on Earth, each with its own unique personality and talents.

Graphite: The Flaky Artist

Graphite is like a stack of paper, with each sheet a honeycomb of carbon atoms. These sheets are loosely bound together, allowing them to slide past each other with ease. This flaky nature makes graphite a fantastic lubricant, like a microscopic ball bearing, reducing friction between moving parts. But graphite has a hidden talent: its free electrons can conduct electricity, making it a star in batteries and electronics.

Diamond: The Unbreakable Superhero

If graphite is the artist, diamond is the superhero. Every carbon atom in diamond is bonded to four others in a rigid, 3D networks. This makes diamond incredibly strong – the hardest natural material known to humankind. It's the ultimate warrior, used in cutting tools and abrasives to conquer the toughest tasks. But beneath its tough exterior lies a beauty: diamond's crystal structure refracts light, creating its signature sparkle.

Silicon Dioxide: The Master of Disguise

Silicon dioxide, or silica, is the chameleon of the group. It's found in many forms, from the sand on the beach to the quartz in your watch. Each silicon atom is bonded to four oxygen atoms, creating a strong, stable structure. This gives silica a high melting point and makes it resistant to most chemicals. But silica is also a master of disguise, able to form glass, ceramics, and even optical fibers that carry information at the speed of light.

Diamond vs. Silica: A Tale of Two Titans

Both diamond and silica are built on a foundation of strong bonds and a tetrahedral arrangement of atoms. They're both hard, have high melting points, and are resistant to chemicals. But they have distinct personalities. Diamond is a pure carbon crystal, while silica is a compound of silicon and oxygen. Diamond is an electrical insulator, while silica can be a semiconductor, making it essential for electronics.

The Power of Giant Covalent Structures

These three materials – graphite, diamond, and silicon dioxide – are just a glimpse into the world of giant covalent structures. Their unique properties, born from the arrangement of their atoms, make them essential for countless applications. From the pencil you write with to the computer you're reading this on, giant covalent structures are the silent heroes of our modern world.

Metallic bonding

Metallic Bonding: A Symphony of Electrons

Imagine a bustling city at night, a dazzling network of lights interconnected and vibrant. That's a metal! But instead of electricity flowing through wires, it's a sea of electrons surging around a rigid grid of positively charged metal ions. This "electron sea" isn't just a pretty picture; it's the secret behind the superpowers of metals.

1. The Dance of Attraction: A Bond Like No Other

Think of each metal atom as a generous giver, tossing its outermost electrons into a shared potluck. These electrons become nomads, wandering freely throughout the metal like partygoers exploring a grand ballroom. Left behind are the metal ions, now positively charged, forming a sturdy framework that holds the party together.

The magic happens when these positive ions and the swirling electron sea meet. They're irresistibly drawn to each other, creating a bond that's strong and flexible, like a troupe of expert dancers moving in perfect harmony. This is the essence of the metallic bond, a dynamic embrace that gives metals their unique character.

Key Characteristics of Metallic Bonding:

Electron Nomads: The valence electrons are free spirits, roaming wherever they please within the metal.
Electrostatic Tango: The bond is a passionate dance between positive ions and the swirling electron sea.
All-Directional Attraction: The attraction is a free-for-all, with forces pulling in every direction, creating a tight-knit community.
Giant Metallic Metropolis: Metals are vast cities of atoms, held together by this intricate dance of electrons and ions.

2. Structure and Bonding: The Metal's Personality Traits

This unique structure and bonding give metals their distinctive personalities. Let's explore some of their most remarkable traits:

(a) Electric Superhighways

Metals are like superhighways for electricity. Those free-roaming electrons are expert couriers, ready to zip towards a positive terminal at a moment's notice, carrying electric current with them. It's like a network of roads with no speed limits or traffic jams! This is why metals are the stars of the electrical world, used in everything from tiny circuits to massive power lines.

(b) Shape-Shifting Masters

Metals are masters of disguise, easily changing shape without falling apart. They can be hammered into thin sheets or stretched into long wires, all thanks to the flexible nature of the metallic bond. When a force is applied, the layers of ions glide smoothly over each other, like a deck of cards being shuffled. The electron sea acts as a lubricant, ensuring the metal stays cohesive even while it's being reshaped.

Other Amazing Metal Abilities:

Toughness: The strong attraction between ions and electrons makes metals incredibly tough, able to withstand high temperatures and resist breaking.
Shining Stars: Metals have a captivating sparkle. This is because the electron sea can capture and release light, creating a mesmerizing shimmer.
Heat Superconductors: Metals are excellent at sharing heat, thanks to those energetic electrons that quickly spread thermal energy throughout the metal.

3. The Strength Within: Factors Influencing the Metallic Bond

The strength of a metallic bond depends on a few key factors:

Electron Abundance: More electrons in the sea mean a stronger, more vibrant bond.
Ionic Charge: A higher positive charge on the metal ions creates a stronger attraction to the electron sea.
Ionic Size: Smaller ions hold onto the electron sea more tightly, leading to a stronger bond.

4. Metal Superstars: Copper, Aluminum, and Iron

Copper: The electrical superstar, used in wires and circuits everywhere. It's also incredibly flexible, making it perfect for pipes and tubes.
Aluminum: The lightweight champion, used in airplanes and other structures where weight matters. It's also a master of disguise, appearing in everything from soda cans to window frames.
Iron: The backbone of civilization, used to create steel, the foundation of buildings, bridges, and countless other structures.

5. Alloy Adventures: Mixing Metals for Superpowers

Alloys are like metal cocktails, blending different elements to create unique combinations. By mixing metals, we can enhance their strengths and create materials with extraordinary properties. Steel, for example, is a powerful blend of iron and carbon, much stronger than iron alone.

6. Metals in Our Modern World: From Smartphones to Skyscrapers

Metals are the silent heroes of our modern world, powering our technologies and shaping our cities. From the tiny components in our smartphones to the massive beams supporting our skyscrapers, metals are essential to our everyday lives.

7. Beyond the Basics: The Quantum Symphony

To truly understand metallic bonding, we need to dive into the quantum world, where electrons behave like waves, creating a mesmerizing symphony of energy. This symphony allows electrons to flow freely, giving metals their incredible conductivity.

8. The Future of Metals: Shaping Tomorrow's World

Scientists are constantly exploring new ways to manipulate metals, creating materials with even more incredible properties. Imagine superconductors that transport electricity with zero resistance, or super-strong, lightweight alloys that revolutionize transportation. The future of metals is bright, filled with endless possibilities.

By understanding and harnessing the power of metallic bonding, we can unlock the full potential of metals and create a future filled with amazing new materials and technologies.

Stoichiometry

Formulae

Unlocking the Secrets of Chemical Formulas: A Journey into the Building Blocks of Matter

Imagine a world where you could understand the language of the universe, where the secrets of matter are laid bare before your eyes. This is the power of chemical formulas – a key to unlocking the mysteries of everything around us, from the air we breathe to the stars in the night sky.

Elements and Compounds: The Characters of Our Chemical Story

Think of elements as the letters of the chemical alphabet, each with its unique personality and characteristics. Hydrogen (H_2), the smallest and most abundant element, dances in pairs like mischievous twins. Oxygen (O_2), another lively duo, gives life to our planet. Nitrogen (N_2) fills the air with a quiet strength, while chlorine (Cl_2) adds a touch of playful mischief.

These elements combine in countless ways to form compounds, the words in our chemical language. Water (H_2O), the elixir of life, unites two hydrogens with a single oxygen. Carbon dioxide (CO_2) exhales from our lungs, a testament to the cycle of life. Sodium chloride ($NaCl$), the salt that flavors our food, is a simple union of sodium and chlorine.

Molecular Formulas: A Precise Portrait of a Molecule

Imagine a molecular portrait, capturing the exact number of each element in a single molecule. This is the power of the molecular formula. It's like a family photo, showing all the members and their relationships.

Water's molecular formula (H_2O) reveals its close-knit family of two hydrogen atoms and one oxygen. Glucose ($C_6H_{12}O_6$), the sugar that fuels our bodies, is a bustling family of six carbons, twelve hydrogens, and six oxygens.

Empirical Formulas: A Simplified Sketch

Sometimes, we need a quick sketch instead of a detailed portrait. The empirical formula gives us the simplest ratio of elements in a compound. It's like a minimalist drawing, capturing the essence of the molecule.

Water's empirical formula remains H2O, its simplicity reflecting its essential nature. Glucose, however, simplifies to CH2O, revealing the basic building block of this complex sugar.

Visualizing Formulas: Building Molecules with Balls and Sticks

Imagine building molecules with colorful balls and sticks, like a child's construction set. This is the world of molecular models, where we can visualize the arrangement of atoms and deduce their formulas.

A model with two black carbon atoms, six white hydrogen atoms, and one red oxygen atom reveals the molecular formula C2H6O. This hands-on approach brings chemistry to life, allowing us to explore the architecture of molecules.

Case Studies: Real-World Applications of Formulas

Let's step out of the lab and into the real world, where formulas play a vital role in understanding the substances around us.

Water vs. Hydrogen Peroxide: Though both contain hydrogen and oxygen, their formulas reveal their distinct personalities. Water (H2O) nurtures life, while hydrogen peroxide (H2O2) has a fiery nature, used for cleaning and disinfecting.

Glucose vs. Fructose: These sugars share the same molecular formula (C6H12O6) but have different arrangements of atoms, like two families with distinct personalities. Glucose provides quick energy, while fructose offers a sweeter taste.

Beyond the Basics: Exploring Advanced Concepts

Our journey into the world of formulas doesn't end here. We can delve deeper, exploring polyatomic ions (groups of atoms with a charge) and hydrates (compounds that hold water molecules). We can even use experimental data to decipher the formulas of unknown substances, like detectives solving a chemical puzzle.

Conclusion: The Language of the Universe

Chemical formulas are more than just symbols and numbers; they are the language of the universe, allowing us to communicate the composition of matter with precision and clarity. By mastering this language, we unlock the secrets of the world around us and gain a deeper appreciation for the beauty and complexity of chemistry.

Imagine this: You're building with LEGOs, but these aren't your ordinary bricks. These are charged LEGOs! Some are positive (cations), some are negative (anions), and they only stick together if the charges match up perfectly. That's how ionic compounds work – they're like tiny LEGO castles built with electrical forces.

Take table salt (NaCl): Sodium (Na+) is a positive brick, chlorine (Cl-) is a negative brick. Snap them together, and voilà! You've got a neutral, stable structure. But what about magnesium chloride? Magnesium (Mg2+) is like a double-positive brick. You'll need two chlorine bricks (Cl-) to balance it out, making MgCl2. It's all about finding the perfect balance!

Now, picture a cooking recipe: You've got your ingredients (reactants) and your finished dish (products). A chemical equation is like that recipe, showing how atoms rearrange themselves in a reaction.

Word equation: "Flour, eggs, and milk bake into a cake"
Symbol equation: It's like using chemical shorthand: Flour + Eggs + Milk → Cake
Ionic equation: This zooms in on the key players in a water solution, like showing how the baking powder reacts to make the cake rise.
And don't forget the state symbols! They're like little labels telling you if something is solid (s), liquid (l), gas (g), or dissolved in water (AQ).

It's the difference between ice, water, steam, and sugar dissolved in your tea!

Balancing equations is like a puzzle: You can't create or destroy atoms, just rearrange them. Think of the classic methane combustion:

CH_4 (g) + O_2 (g) → CO_2 (g) + H_2O (g)

It's like saying, "One methane molecule and two oxygen molecules combine to make one carbon dioxide molecule and two water molecules." Gotta makes sure all the atoms are accounted for on both sides!

Let's get creative! Imagine you're a detective at a chemical crime scene. You find some clues (information about a reaction) and have to deduce the symbol equation. It's like piecing together the evidence to figure out what happened!

For example: You know silver nitrate and sodium chloride react to form a white solid (silver chloride). With your detective skills, you deduce:

$AgNO_3$ (AQ) + NaCl (AQ) → AgCl (s) + $NaNO_3$ (AQ)

You've cracked the case!

From batteries to fertilizers, these concepts are everywhere! Understanding ionic compounds and chemical equations is like having a secret decoder ring for the universe. It helps us understand how everything works, from the smallest atom to the largest star. So, embrace your inner chemist and unlock the wonders of the world around you!

Relative masses of atoms and molecules

Imagine the tiniest of scales, so small it could weigh a single atom! That's the world of relative atomic mass (AR) and relative molecular mass (Mr.). We're talking about the masses of atoms and molecules, things so incredibly tiny that grams and kilograms just won't cut it.

1. Relative Atomic Mass (AR): The Atomic Popularity Contest

Think of it like a popularity contest for atoms, where the "coolest" kid, carbon-12 (^{12}C), sets the standard. Everyone else's popularity (or mass) is measured relative to 1/12th of carbon-12's mass.

But here's the twist: most elements come in different "flavors" called isotopes. These isotopes are like siblings with the same number of protons but different numbers of neutrons. So, they have the same atomic number but different masses.

To find the relative atomic mass of an element, we take a weighted average of all its isotopes, like calculating the average popularity of all the siblings in a family. The more abundant an isotope, the more it influences the final "popularity" score.

Why does AR matter?

Molar Mass: AR, expressed in grams, tells us the mass of one mole of an element, which is like a chemist's "dozen" of atoms.
Stoichiometry: It's the key to understanding chemical recipes (reactions) and figuring out how much of each ingredient (reactant) is needed to make a certain amount of product.
Elemental Personality: AR gives us clues about an element's physical and chemical properties, like its melting point or how it reacts with other elements.

2. Relative Molecular Mass (Mr.): The Molecular Weightlifting Competition

Now, imagine a weightlifting competition for molecules. Mr. is like the total weight lifted by a team of atoms in a molecule. For ionic compounds, which are like tightly-knit communities of atoms, we use the term "relative formula mass."

How do we calculate Mr.?

It's simple! Just add up the relative atomic masses of all the atoms in the molecule. Think of it as adding up the individual weights lifted by each member of the weightlifting team.

Why is Mr. important?

Molar Mass: Mr., expressed in grams, gives us the mass of one mole of a compound.
Stoichiometry: It's essential for understanding chemical reactions involving compounds, like figuring out how much cake you can bake with a certain amount of flour and sugar.
Chemical Detective Work: Mr. helps us identify unknown compounds and determine their molecular formulas.

Real-World Applications:

Pharmaceuticals: Mr. is crucial for developing new drugs, calculating dosages, and ensuring drug safety.
Environmental Science: It helps us analyze pollutants, monitor water quality, and understand environmental changes.
Materials Science: Mr. is used to design new materials with specific properties, like stronger plastics or lighter alloys.

In a nutshell:

Relative atomic mass (AR) and relative molecular mass (Mr.) are like tiny scales and weightlifting competitions in the world of chemistry. They help us understand the masses of atoms and molecules, which is crucial for everything from developing new medicines to protecting the environment.

Imagine this:

You're a detective investigating a mysterious substance. You need to know its identity, how much space it takes up, and how potent it is. Your tools? The mole, molar gas volume, and concentration!

1. The Mole: Your Tiny Detective's Magnifying Glass

Think of the mole as your super-powered magnifying glass. It allows you to "see" and count the tiniest particles – atoms, molecules, ions – the building blocks of everything around us.

But these particles are so small, trying to count them individually would be like counting grains of sand on all the beaches in the world! That's where Avogadro's constant comes in. It's like a magical shortcut, telling you that one mole of anything contains 6.02×10^{23} particles.

Example:

You've found a mysterious powder at a crime scene. It's 10 grams of pure carbon. Your mole magnifying glass reveals that it contains a mind-boggling 5.01×10^{23} carbon atoms! That's a lot of suspects!

2. Molar Gas Volume: The Case of the Expanding Gas

Now, imagine you're dealing with a gas – a substance that loves to spread out and fill any space it finds. How do you measure something so elusive?

Enter molar gas volume! It's like a special container at room temperature and pressure that always holds exactly one mole of any gas, taking up about 24 liters of space.

Example:

You're analyzing a gas sample collected from a volcanic eruption. You find it occupies 12 liters at room temperature. Using your molar gas volume "container," you deduce that you have 0.5 moles of this gas. Now, you can move on to identifying the culprit behind the eruption!

3. Concentration: The Strength of the Potion

Imagine you're brewing a magic potion. You need to know how strong to make it – how much of the magical ingredient is dissolved in the liquid. That's where concentration comes in.

It tells you how "crowded" the potion's active ingredient is within the liquid. You can measure this in grams per liter (g/dm^3) or moles per liter (mol/dm^3).

Example:

You're creating a healing potion. You dissolve 5 grams of a rare herb in 250 ml of water. Your concentration calculations reveal that your potion has a strength of 20 g/dm^3. Just the right amount to cure a common cold, but not strong enough to cause side effects!

Solving the Mystery

By mastering the mole, molar gas volume, and concentration, you become a true "substance detective." You can unravel the secrets of matter, from the tiniest atom to the largest cloud of gas.

So, embrace your inner detective, grab your tools, and get ready to explore the exciting world of chemistry!

Imagine this: You're holding a bag of sugar. Seems simple, right? But inside that bag, there's a mind-bogglingly huge number of tiny sugar molecules, all crammed together. Like, trillions upon trillions of them. How do we even begin to make sense of such enormous numbers?

That's where the mole comes in. It's like a magical counting tool for chemists. Instead of counting individual molecules (which would take forever!), we count them in groups called moles. Think of it like a dozen eggs – you wouldn't count each individual egg, you'd just say "one dozen."

The Mole's Secret Code:

The mole is all about connecting the stuff we see and touch (like that bag of sugar) to the invisible world of atoms and molecules. And the key to this connection is a special equation:

amount of substance (mol) = mass (g) / molar mass (g/mol)

This equation is like a secret code that unlocks a whole bunch of chemistry puzzles! It lets us figure out:

How many moles are in a certain amount of stuff? (Like, how many moles of sugar are in your bag?)
How much does a certain number of moles weigh? (If you have 0.5 moles of water, how many grams is that?)
What's the molar mass of a substance? (This tells us how much one mole of something weighs.)

Let's Play with the Mole!

1. Counting Moles:

Imagine you have 10 grams of water. How many moles of water is that? First, we need to know the molar mass of water, which is like its "molecular weight." We find that by adding up the atomic weights of the hydrogen and oxygen atoms in water. Once we have the molar mass, we can use our secret code equation to find the number of moles.

2. Weighing Moles:

Let's say you have 0.25 moles of carbon dioxide (the stuff we breathe out). How much does that weigh? Again, we use our equation, this time plugging in the number of moles and the molar mass of carbon dioxide.

3. Unmasking the Molar Mass:

You have a mysterious substance, and you know that 5 moles of it weighs 115 grams. What's its molar mass? Our trusty equation helps us crack this case too!

4. Relative Atomic Mass: The Mole's Cousin:

The mole has a close relative called the relative atomic mass. It's basically a way of comparing the masses of different atoms. We use it to calculate the molar mass of molecules.

5. Counting Particles: The Mole's Superpower:

Remember those trillions of sugar molecules? The mole has the superpower to count them! It uses something called Avogadro's constant, which is a giant number that tells us how many particles are in one mole.

The Mole in Action: A Chemical Adventure!

The mole isn't just about calculations – it's a key player in understanding how chemical reactions work. For example, let's say we're burning methane (the main ingredient in natural gas). We can use the mole to figure out exactly how much carbon dioxide will be produced.

The mole is your guide to the fascinating world of chemistry! It helps us understand the hidden connections between the things we see and the tiny particles that make up everything around us. So, embrace the mole, and let it led you on a journey of discovery!

1. Stoichiometry: The Recipe of Chemistry

Imagine you're a chef, but instead of cooking delicious dishes, you're creating amazing chemical reactions! Stoichiometry is like your recipe book, guiding you on the exact amounts of ingredients (reactants) needed to create your desired products.

Reacting Masses: The Balancing Act

Just like a chef needs to balance flavors, a chemist needs to balance equations. This ensures you have the perfect ratio of reactants to avoid any "leftovers" (excess reactants) or "burnt dishes" (incomplete reactions).

Moles: The Chemical Counting Units

Instead of cups and teaspoons, chemists use moles. Think of moles as tiny measuring scoops for atoms and molecules.

Example: Baking Soda Volcano

Remember those erupting volcanoes from science class? That's stoichiometry in action! The balanced equation tells you exactly how much baking soda and vinegar to mix for the perfect eruption.

Limiting Reactants: The Party Pooper

Sometimes, you run out of one ingredient before the others. This is your limiting reactant – the party pooper that stops the reaction from going further.

Example: Campfire S'mores

Imagine you have enough graham crackers and chocolate but only a few marshmallows. Your marshmallows are the limiting reactant, limiting the number of s'mores you can make.

Volumes of Gases: Balloons and Bubbles

Gases are like mischievous balloons, always wanting to expand and fill space. At room temperature and pressure (R.T.P.), one mole of any gas takes up about 24 dm3 – that's like a big box full of gas!

Example: Blowing Bubbles

When you blow bubbles, you're creating gas-filled spheres. The amount of soap solution you use determines the number of moles of air you can trap, and thus the number of bubbles you can make.

Volumes of Solutions: The Potion Master

Solutions are like magical potions, with different concentrations of ingredients.

Example: Making Lemonade
If you like your lemonade strong (high concentration), you'll use more lemon juice and sugar in a given amount of water.

Concentrations of Solutions: The Flavor Test

Concentration tells you how strong your solution is.

Example: Taste Testing
Imagine tasting different salt solutions. The one that tastes the saltiest has the highest concentration of salt.

Conversion between cm3 and dm3: Metric Magic

Converting between cubic centimeters (cm3) and cubic decimeters (dm3) is like using magic spells to change the size of your containers.

2. Titration: The Chemical Detective

Titration is like a detective game, where you slowly add one solution to another until you reach the "crime scene" – the endpoint of the reaction. This helps you determine the unknown concentration of a solution.

Example: Finding the Culprit
Imagine you're a detective trying to identify a mystery liquid. By carefully adding a known solution (your "detective tool"), you can figure out the exact nature of the unknown liquid.

3. Empirical and Molecular Formulas: The Chemical Code Breakers

Empirical and molecular formulas are like secret codes that reveal the identity of a compound.

Empirical Formula: The Simplified Code

This gives you the simplest ratio of atoms in a compound.

Molecular Formula: The Full Code
This tells you the actual number of each type of atom in a molecule.

Example: Deciphering a Message
Imagine receiving a coded message. The empirical formula gives you the basic building blocks of the code, while the molecular formula reveals the complete message.

Remember: Chemistry is all around us, from the food we eat to the air we breathe. By understanding stoichiometry, you can unlock the secrets of the chemical world and appreciate the magic happening all around you!

Electrochemistry

Electrolysis

Zapping Molecules Back to the Stone Age: An Electrifying Tale of Electrolysis!

Imagine a microscopic courtroom drama. On one side, we have the anions, the negative ions, nervously shuffling towards the anode, the positive electrode. On the other, the cations, positive ions, confidently strutting towards the cathode, the negative electrode. The judge, a powerful electrical current, slams their gavel (or should we say, battery?) and shouts, "Order in the court! Today, we're breaking bonds and forging new identities!"

This, my friends, is the electrifying world of electrolysis – where we use electricity to split molecules back into their elemental building blocks, like a cosmic reset button. It's like taking a LEGO castle and zapping it back into a pile of bricks, ready to be built into something new.

The Players:

Electrolytic Cell: Our courtroom stage, a container where the drama unfolds.
Electrolyte: The witness stand, a liquid or molten substance teeming with ions ready for questioning (or should we say, splitting?).
Electrodes: The jury, two conductors (often metals) that deliver the judge's verdict (the electric current).

The Plot:

Ionization: The molecules enter the courtroom and swear to tell the truth, the whole truth, and nothing but the truth, so help them electricity! They split into their charged components – the ions.
Migration: Like moths to a flame, the anions are drawn to the positive anode, while the cations head towards the negative cathode.

Discharge: At the electrodes, the ions face their judgment. They either gain or lose electrons, transforming into neutral atoms or new molecules.
Product Formation: The transformed particles exit the courtroom, some settling down as solids on the electrodes, others bubbling away as gases, and some remaining dissolved in the electrolyte.

Why Should We Care?

Electrolysis isn't just a microscopic soap opera; it's a powerful tool with real-world applications:

Metal Extraction: Think aluminum cans, shiny jewelry, and strong steel beams – electrolysis helps us extract these metals from their ores.
Purification: Electrolysis refines metals like copper, making them purer and more valuable.
Electroplating: Ever wondered how your cheap jewelry gets that golden sheen? Electroplating! It deposits a thin layer of metal onto another, for decoration or protection.
Chemical Production: From chlorine used in swimming pools to sodium hydroxide used in soap, electrolysis is a key player in the chemical industry.
Clean Energy: Electrolysis can split water into hydrogen and oxygen, providing clean fuel for the future.

The Future of Zapping:

Scientists are constantly pushing the boundaries of electrolysis, developing new techniques like:

Proton Exchange Membrane (PEM) Electrolysis: A super-efficient way to produce hydrogen.
Solid Oxide Electrolysis Cells (SOECs): High-temperature electrolysis that can work with renewable energy sources.
Microbial Electrolysis Cells (MECs): Using tiny microbes to help with electrolysis, especially for cleaning wastewater.
Electrolysis is a testament to our ability to harness the power of electricity to transform matter. It's a process that's shaping our world, from the everyday objects we use to the clean energy technologies of the future. So next time you see a shiny chrome bumper or sip a

refreshing soda from an aluminum can, remember the tiny courtroom drama of electrolysis that made it all possible!

Imagine a bustling city:

Instead of diving straight into technical terms, let's start with a relatable analogy. Imagine a bustling city with two distinct districts: Anode Avenue, known for its excess of goods (electrons), and Cathode Corner, facing a shortage. To maintain balance, a clever network of roads (wires) and a powerful central hub (the battery) facilitate the transfer of goods.

Electrons on the move:

Think of electrons as tiny delivery trucks zipping along these roads, carrying precious cargo from Anode Avenue to Cathode Corner. The battery acts like a central dispatcher, ensuring a constant flow of these trucks to keep the city running smoothly. This flow is what we call "electric current," the lifeblood of our electrochemical city.

Electrodes: The Transformation Hubs:

At the heart of each district lie bustling transformation hubs – the electrodes. Anode Avenue's hub specializes in taking apart bulky goods (ions) and extracting valuable components (electrons). Cathode Corner's hub, on the other hand, expertly assembles these components into new and useful products.

Ions: The Raw Materials:

Now, let's introduce the "ions" – the raw materials that keep these hubs humming. Positive ions, or "cations," are drawn to Cathode Corner's assembly hub, eager to gain electrons and transform. Negative ions, or "anions," head towards Anode Avenue's extraction hub, ready to shed their excess electrons.

Electrolyte: The City's Transportation System:

Connecting these districts is a sophisticated transportation system – the electrolyte. It could be a molten highway or a solution-filled canal, allowing ions to smoothly navigate between the transformation hubs.

Case Studies: Real-World Applications:

Sodium and Chlorine: A City Divided:

Imagine a city where sodium and chlorine, once tightly bound, are separated through electrolysis. Anode Avenue extracts electrons from chlorine, releasing it as a gas, while Cathode Corner welcomes sodium ions, transforming them into pure metal.

Copper Plating: A City Beautification Project:

Think of electroplating as a city beautification project. A copper object, dull and worn, is placed in Cathode Corner. The hub uses copper ions from the electrolyte and electrons from the battery to give the object a shiny new coat, enhancing its appearance and value.

Conclusion: The city that Never Sleeps:

This electrochemical city, powered by the ceaseless flow of electrons and ions, is a microcosm of the electrolysis process. By understanding the roles of its different components, we can appreciate the intricate dance of charge transfer that drives this essential process.

This humanized approach, filled with relatable analogies and vivid imagery, aims to make the concept of charge transfer during electrolysis more accessible and engaging. It encourages readers to visualize the process, fostering a deeper understanding and appreciation for this fundamental electrochemical phenomenon.

Imagine a microscopic party where molecules are breaking up and swapping partners! That's kind of what electrolysis is like. It's a chemical drama where we use electricity to split compounds into their elements. Think of it as a forced divorce for molecules!

Here's what you need for this molecular break-up:

The Dance Floor: A container filled with a substance that has free ions (charged particles). This is our "electrolyte," and it can be a molten ionic compound or an ionic compound dissolved in water.
The Matchmakers: Two electrodes, usually made of unreactive materials like platinum or carbon, dipped into the electrolyte. One's connected to the positive terminal of a battery (the anode), and the other to the negative terminal (the cathode). They act like magnets, attracting the ions.

Now, let's get this party started!

When electricity flows, the positive ions (cations) rush to the cathode for a hookup with electrons, getting reduced. The negative ions (anions) head to the anode to lose electrons and get oxidized. It's a wild exchange of electrons!

Let's check out the specific cases you mentioned:

(a) Molten Lead (II) Bromide (PB Br_2) - A Heavy Metal Breakup

The Scene: Molten lead (II) bromide is like a hot, swirling dance floor with lead (Pb^{2+}) and bromide (Br^-) ions moving freely.
The Action:
At the cathode, lead ions grab electrons and turn into pure lead metal. It's like they've suddenly cooled down and settled at the bottom.
At the anode, bromide ions lose electrons and pair up to form bromine gas, a reddish-brown fume that escapes the party.

(b) Concentrated Aqueous Sodium Chloride (NaCl) - A Salty Affair

The Scene: A concentrated saltwater solution is a crowded dance floor with sodium (Na^+), chloride (Cl^-), and water molecules all mingling.
The Action:
At the cathode, water molecules steal the show and grab electrons, forming hydrogen gas bubbles and leaving behind hydroxide ions. It's like they crashed the party and kicked sodium out!

At the anode, chloride ions lose electrons and form chlorine gas, a pale green-yellow gas with a bleaching effect. It's like they got so excited they started cleaning up the place!

(c) Dilute Sulfuric Acid (H_2SO_4) - An Acidic Split

The Scene: Dilute sulfuric acid is a more intimate gathering with hydrogen (H^+) ions, sulfate (SO_4^{2-}) ions, and water molecules.
The Action:
At the cathode, hydrogen ions gain electrons and form hydrogen gas bubbles.
At the anode, water molecules lose electrons, forming oxygen gas bubbles and more hydrogen ions. It's like a never-ending cycle!

Remember these key points:

Inert Electrodes: We use unreactive electrodes so they don't join the party and mess things up.
Concentration Matters: The concentration of the electrolyte can change the outcome of the party.
Overpotential: Sometimes, we need to give the party a little extra "push" (voltage) to get things going.

Electrolysis is used in tons of applications:

Extracting Metals: Like getting aluminum from its ore.
Purifying Metals: Like refining copper.
Producing Chemicals: Like making chlorine and sodium hydroxide.
Electroplating: Like coating objects with a thin layer of metal.
So, there you have it! Electrolysis is a fascinating process with a wide range of applications. It's like a molecular mixer where we can use electricity to break up and create new substances.

Imagine a pool party... but for charged particles!

Our main guests are the copper (II) ions (Cu^{2+}), looking dapper in their blue outfits, and the sulfate ions (SO_4^{2-}), their stylish partners. They're all swimming around in a luxurious copper (II) sulfate solution, enjoying the good life.

Suddenly, we introduce two giant party crashers – our electrodes!

Scenario 1: The 'Wallflower' Electrodes (Inert)

These electrodes, made of boring old carbon or graphite, just want to hang back and watch the action. They don't like to get involved.

At the Cathode (the Negative Nancy): The Cu^{2+} ions, feeling a bit shy and attracted to the negative vibes, gather around the cathode. They grab some electrons from the electrode (like grabbing a drink at the bar) and transform into solid copper (Cu). Think of it as them getting a bit "grounded" and settling down. You'll see a reddish-brown layer forming on the electrode – that's our copper party crashers!
At the Anode (the Positive Pollyanna): The OH^- ions from the water, always up for a good time, flock to the anode. Feeling energized by the positive vibes, they lose some electrons (like hitting the dance floor and letting loose!). This turns them into oxygen gas (O_2) and water (H_2O). You'll see bubbles forming – that's the oxygen getting the party started!
The Aftermath: The pool party starts to lose its blue hue as the Cu^{2+} ions disappear, leaving the solution a bit dull.

Scenario 2: The 'Life of the Party' Electrodes (Active)

This time, the electrodes are made of copper themselves – they're ready to mingle!

At the Cathode (still a Negative Nancy): Same as before, the Cu^{2+} ions are drawn to the negative vibes and grab some electrons, transforming into solid copper. They basically join the electrode and make it even bigger!
At the Anode (now a Copper Casanova): This time, the copper electrode itself gets in on the action. Copper atoms from the electrode lose electrons and become Cu^{2+} ions, jumping into the solution to replace the ones that left. It's like a revolving door of copper ions!
The Aftermath: The pool party stays lively and blue! The concentration of Cu^{2+} ions remain constant, with the anode sacrificing itself to keep the party going.

General Rules to Remember:

Cathodes are like magnets for positive ions (cations). Metals and hydrogen love to hang out there and get reduced (gain electrons). Anodes attract negative ions (anions). Non-metals (except hydrogen) are drawn to the positive energy and get oxidized (lose electrons).

Electrolysis: Not Just a Party Trick!

This process has real-world applications, like:

Electroplating: Giving cheap jewelry a fancy gold or silver coating.
Producing Aluminum: Making those lightweight cans for your soda.
Purifying Copper: Turning dirty copper into shiny, usable metal.

Safety First, Party Animals!

Remember, electrolysis can be dangerous if you're not careful. Always wear your safety goggles and gloves, and make sure there's good ventilation to avoid breathing in any nasty gases.

So, there you have it! Electrolysis, explained with a dash of creativity and a sprinkle of fun. Hopefully, this "humanized" version helps you understand the process better. Now go forth and electrify your mind!

Electrolysis and Electroplating: A Magical Metal Makeover

Imagine a world where you could transform a rusty old spoon into a gleaming piece of silverware with the flick of a switch. That's the magic of electroplating! But first, let's dive into the fascinating world of electrolysis, the process that makes this transformation possible.

1. Electrolysis: Splitting Molecules with Electricity

Think of electrolysis like a molecular divorce court. We use electricity to break up stable couples (like halide compounds) and force them to mingle with other singles at a wild electrochemical party.

The Party Venue: Our party takes place in an electrolytic cell, a special container with two electrodes: the anode (+) and the cathode (-).

The Guests: Halide compounds (like table salt) dissolve in water, splitting into positive and negative ions. Water molecules also join the party, bringing their own positive and negative ions.
The Music: When we turn on the electric current, it's like turning up the music and getting the party started! The ions start moving and grooving towards the electrode with the opposite charge.
The Matchmaking: At the electrodes, the ions meet their match and undergo chemical reactions.
At the anode (+), negative ions lose electrons and form new substances (like chlorine gas).
At the cathode (-), positive ions gain electrons and can even transform into pure metals (like shiny copper)!

The "Who's Who" of Electrolysis:

Reactivity Series: This is like a popularity contest for elements. The more reactive an element, the more likely it is to steal electrons and form new bonds.
Concentration: Imagine a crowded dance floor versus a more intimate setting. In concentrated solutions, halide ions are more likely to react, while in dilute solutions, water molecules get a chance to shine.

Example: The Seawater Spa Treatment

Seawater is like a giant electrolyte cocktail. When we use electrolysis on seawater, we can produce chlorine gas (used in swimming pools and cleaning products) and hydrogen gas (a potential fuel source). It's like giving the ocean a spa treatment and getting valuable products in return!

2. Ionic Half-Equations: The Chemical Gossip

Half-equations are like the gossip columns of the electrolysis world. They tell us exactly what's happening at each electrode.

For example, when we electrolyze a concentrated solution of table salt (sodium chloride):

Anode Gossip: $2Cl^-(AQ) \rightarrow Cl_2(g) + 2e^-$ (The chloride ions lose electrons and form chlorine gas - drama!)

Cathode Gossip: $2H_2O(l) + 2e^- \rightarrow H_2(g) + 2OH^-(AQ)$ (Water molecules gain electrons and form hydrogen gas - a new romance!)

3. Electroplating: The Ultimate Metal Makeover

Electroplating is like giving an object a metallic makeover. We use electrolysis to coat it with a thin layer of a more attractive or durable metal.

Benefits of Electroplating:

Beauty Boost: Turn a plain object into a dazzling masterpiece by coating it with gold, silver, or even colorful metals.
Anti-Aging Treatment: Protect metals from corrosion (rust) and keep them looking young and fresh.
Strength Training: Make surfaces more durable and resistant to scratches and wear.
Conductivity Enhancement: Improve electrical conductivity for electronic applications.

4. The Electroplating Process: A Step-by-Step Guide

Prep the Canvas: Clean the object thoroughly to ensure the new metal layer adheres properly.
Set the Stage: Create an electrolytic cell with the object as the cathode (-) and the plating metal as the anode (+).
Mix the Potion: Immerse both electrodes in an electrolyte solution containing ions of the plating metal.
Cast the Spell: Turn on the electric current and watch as metal ions from the solution magically deposit onto the object, creating a stunning new look.
Finishing Touches: Rinse, dry, and polish the electroplated object to perfection.

Examples of Electroplating Magic:

Jewelry: Give your favorite pieces a luxurious upgrade with gold or silver plating.
Automotive Parts: Protect car parts from rust and give them a sleek, shiny finish with chromium plating.

Electronics: Ensure reliable electrical connections with gold-plated contacts.
Cutlery: Prevent tarnishing and add a touch of elegance to your dining table with silver-plated cutlery.

The Future of Electroplating: More Magic to Come!

Scientists are constantly developing new electroplating techniques to create even more amazing transformations:

Pulse Plating: Using a pulsed electric current to create stronger and more uniform coatings.
Nanoparticle Electroplating: Adding tiny nanoparticles to enhance the properties of the coating, like making it super strong or resistant to scratches.
Alloy Electroplating: Creating coatings with unique properties by mixing different metals.

Conclusion:

Electrolysis and electroplating are powerful tools that allow us to manipulate and transform materials in incredible ways. From producing essential chemicals to creating beautiful and durable objects, these processes play a vital role in our modern world. So next time you admire a shiny piece of jewelry or a gleaming car bumper, remember the magic of electrolysis and electroplating that made it possible!

Hydrogen-oxygen fuel cells

Fuel Cells: Where Science Meets Sustainability (and maybe, just maybe, saves the planet!)

Imagine a car that runs on air and water. Sounds like something out of a sci-fi movie, right? Well, buckle up, because hydrogen-oxygen fuel cells are bringing that fantasy to life!

Think of it like this: these ingenious contraptions are like tiny power plants, humming away under the hood of your car. They take hydrogen (the most abundant element in the universe!) and oxygen (that stuff we breathe) and, through some seriously cool chemistry, transform them into electricity. The only exhaust? Pure, clean water vapor. Mother Nature approves!

Why Fuel Cells are the Future's Hottest Ticket:

Goodbye, Smog City! Say "adios" to those nasty greenhouse gases and air pollutants. Fuel cells are zero-emission heroes, making our cities cleaner and our lungs happier.
Efficiency is King: Forget those gas-guzzling dinosaurs! Fuel cells squeeze way more energy out of their fuel, meaning you can go farther on less. Your wallet will thank you.
Silence is Golden: No more roaring engines! Fuel cells operate with a whisper, making your ride smooth, serene, and oh-so-stylish.
Renewable Powerhouse: Hydrogen can be made from renewable sources like solar and wind power. Talk about a sustainable energy dream team!
Refuel in a Flash: Filling up a hydrogen car is as quick and easy as a regular gas station pit stop. No more waiting around for hours while your battery charges!
Road Trip Ready: Fuel cell cars can go the distance, rivaling even the best electric vehicles in range. Cross-country adventures, here we come!

Okay, Okay, There Are a Few Hiccups:

Hydrogen Highways: We need to build a whole new infrastructure to produce and distribute hydrogen. It's like building a network of roads, but for this amazing new fuel.
Storage Solutions: Hydrogen is a bit of a wild child, so we need special high-tech tanks to keep it safely contained.
Price Tag: Fuel cells are still a bit pricey, but like all cool new tech, the cost will come down as it becomes more common.
Energy Efficiency: While fuel cells are super-efficient, there are still some energy losses during hydrogen production and transportation. We're working on it!
Public Perception: Many people haven't even heard of fuel cells! We need to spread the word about this awesome technology.

Let's Get Technical (Just a Little):

Hydrogen Production: We can make hydrogen from natural gas, water, or even biomass! Scientists are constantly finding new and cleaner ways to produce it.
Fuel Cell Types: There are different types of fuel cells, each with its own strengths. Proton exchange membrane fuel cells (PEMFCs) are the stars of the show for vehicles.
Fuel Cell Pioneers: Companies like Toyota, Hyundai, and Honda are already rolling out impressive fuel cell cars like the Mirai, NEXO, and Clarity.
Real-World Success: California is leading the way with fuel cell adoption, and countries like Germany and Japan are hot on their heels.

The Future is Fuel Cells!

Hydrogen fuel cell technology is a game-changer. It has the potential to revolutionize transportation and create a cleaner, greener world. While challenges remain, the future looks bright. With continued innovation and investment, we're on the road to a sustainable future powered by hydrogen.

So, the next time you see a car silently gliding by, leaving only a trail of water vapor, remember: that's the future, and it's powered by fuel cells!

Chemical energetics

Exothermic and endothermic reactions

A Tale of Two Reactions: When Molecules Play Hot and Cold

Imagine the world of molecules as a bustling dance floor. Some couples whirl and twirl, radiating energy and heat as they move – these are our exothermic reactions. Others, like wallflowers, draw energy from the room, cooling it down as they sway gently – these are the endothermic reactions.

Exothermic Reactions: The Life of the Party

Think of a roaring bonfire on a chilly night. The crackling flames release waves of warmth, a cozy embrace against the cold. This is the essence of an exothermic reaction – a molecular fiesta where energy is liberated in the form of heat, light, or even sound!

These reactions are the "givers" of the chemical world, generously sharing their energy with their surroundings. They're the reason your hands warm up when you hold a hot cup of cocoa, and why a firework explodes in a dazzling burst of color.

Examples of Exothermic Extravaganzas:

Burning: Whether it's a candle flame or a forest fire, combustion reactions unleash a fiery spectacle of heat and light.
Explosions: From dynamite to the rapid expansion of popcorn, explosions are dramatic displays of energy release.
Neutralization: When acids and bases meet, they neutralize each other, releasing heat and sometimes forming salts.

Endothermic Reactions: The Cool Kids on the Block

Now picture an ice pack soothing a sore muscle. It feels cool to the touch because it's absorbing heat from your body. This is the hallmark of an endothermic reaction – a molecular chill-out session where energy is drawn in from the surroundings.

These reactions are the "takers" of the chemical world, borrowing energy to fuel their transformations. They're the reason a cold compress provides relief, and why baking soda absorbs odors.

Examples of Endothermic Enchantments:

Melting: When ice transforms into water, it needs to absorb heat to break the bonds holding the water molecules in a rigid structure.
Evaporation: Water disappearing into thin air? It's absorbing heat to escape as a gas.
Photosynthesis: Plants are masters of endothermic reactions, using sunlight to convert carbon dioxide and water into energy-rich sugars.

Enthalpy Change (ΔH): The Energy Scorecard

Every chemical reaction has an energy story to tell, and enthalpy change (ΔH) is its narrator. It's the scorekeeper of heat absorbed or released during a reaction, a measure of the energy drama unfolding at the molecular level.

$\Delta H < 0$ (negative): Exothermic reaction – energy is released, the score favors the products.
$\Delta H > 0$ (positive): Endothermic reaction – energy is absorbed, the score favors the reactants.

The Bigger Picture

Understanding the interplay of exothermic and endothermic reactions is like having a backstage pass to the molecular world. It's a key to unlocking the secrets of everything from everyday phenomena to cutting-edge scientific advancements.

So, the next time you strike a match or feel the chill of an ice pack, remember the invisible dance of molecules, playing hot and cold in the fascinating world of chemistry.

Imagine a Chemical Reaction as a Mountain Hike

Picture yourself embarking on a challenging mountain hike. You're full of energy and eager to reach the summit, but a steep incline stands in your way. This incline is like the activation energy in a chemical reaction – the initial energy investment needed to get things going.

Just as you need to exert effort to climb the hill, molecules require energy to break their existing bonds and form new ones. This energy barrier prevents reactions from happening spontaneously all the time, which is a good thing, otherwise, everything around us would be constantly reacting!

Why the Need for This Energy Hurdle?

Think of it like this:

Breaking Up is Hard to Do: Molecules are comfortable in their current bonds. To break these bonds, you need to put in some energy, just like it takes effort to end a relationship.
Opposites Attract (and repel): When molecules get close, their electron clouds can initially repel each other like magnets with the same poles facing each other. Overcoming this repulsion requires energy, like pushing those magnets together.
The Peak of Instability: At the top of the hill, you reach a point of maximum potential energy before you start descending. Similarly, molecules need to reach a high-energy, unstable state called the transition state, where old bonds are breaking, and new ones are forming.

Factors Affecting the Energy Barrier:

The "Type" of Climbers: Some hikers are seasoned pros, while others are beginners. Similarly, reactions involving ions (atoms with a charge) tend to have lower activation energies than those with covalent bonds (shared electrons), because ionic bonds are already partially broken due to their electrostatic attraction.
A Sunny Day for a Hike: A warm, sunny day makes hiking easier. Likewise, increasing temperature gives molecules more kinetic energy, making them more likely to overcome the activation energy barrier.
A Shortcut Through the Mountain: Imagine finding a tunnel through the mountain that makes your climb much easier. This is what a catalyst does! It provides an alternative reaction pathway with a lower

activation energy, speeding up the reaction without being consumed in the process.

Mapping Your Chemical Journey:

Now, imagine having a map that shows the elevation changes throughout your hike. This is like a reaction pathway diagram, which visually represents the energy changes during a chemical reaction.

The Trail: The horizontal axis represents the progress of the reaction, like the trail on your map.
Altitude: The vertical axis shows the energy of the system, similar to the elevation on your map.
Starting Point and Destination: The reactants are your starting point at the base of the mountain, and the products are your destination at the summit or valley.
The Peak: The highest point on the diagram is the transition state, that unstable point at the top of the hill.
Effort Required: The difference in elevation between your starting point and the peak is the activation energy.
Net Change: The overall elevation changes from start to finish represents the enthalpy change (ΔH) – whether you ended up higher or lower than you started.

Types of Hikes (Reactions):

Downhill Hike (Exothermic Reaction): You end up at a lower elevation than where you started, releasing energy like you would by jogging downhill. This released energy can be used for other purposes, like generating heat.
Uphill Hike (Endothermic Reaction): You need to put in a lot of effort to reach a higher elevation, absorbing energy like you would by climbing uphill.

Real-World Examples:

Enzymes: The Body's Hiking Guides: Enzymes are like experienced guides that help your body's chemical reactions happen more easily. They lower the activation energy, allowing reactions to occur at normal body temperatures.

Haber Process: Industrial-Scale Fertilizer Production: The Haber process, used to produce ammonia for fertilizers, is like a massive mountain climbing expedition. It requires a catalyst (like specialized equipment) to make the process feasible on a large scale.
Ozone Depletion: An Unwanted Shortcut: Chlorofluorocarbons (CFCs) act like a harmful shortcut that accelerates the destruction of the ozone layer, our planet's protective shield against harmful UV radiation.

In Conclusion:

Understanding activation energy and reaction pathway diagrams is like having a map and a guide for the chemical world. They help us visualize the energy changes involved in reactions, understand how catalysts work, and appreciate the delicate balance of energy in our everyday lives.

Bond Breaking and Bond Making: A Tale of Two Atoms

Imagine two atoms, let's call them Romeo and Juliet. They're deeply in love and bound together by an invisible force - a chemical bond. But their families (the surrounding molecules) disapprove! To tear them apart, the families need to exert energy, like pulling two magnets apart. This energy is absorbed from the surroundings, making the breakup (bond breaking) an endothermic process. Poor Romeo and Juliet!

But wait! Romeo and Juliet refuse to be separated. They defy their families and come back together in a passionate embrace! This reunion releases energy, like a burst of heat and a sigh of relief. This is bond making, an exothermic process. The energy released warms the hearts of everyone around them (the surroundings).

Enthalpy Change: The Love-O-Meter

The enthalpy change (ΔH) of a reaction is like a love-o-meter, measuring the overall energy change in this atomic romance. If more energy is released when Romeo and Juliet reunite (bond formation) than was needed to tear them apart (bond breaking), the reaction is exothermic (ΔH is negative). Love conquers all!

But if it took more energy to break them up than they release when they get back together, the reaction is endothermic (ΔH is positive). Alas, sometimes love isn't enough to overcome the challenges.

Calculating Enthalpy Change: Counting the Cost of Love

Every bond has a certain strength, measured by its bond energy. Think of it as the effort required to break up the lovebirds. We can use bond energies to estimate the enthalpy change of a reaction:

$\Delta H = \Sigma$ (bond energies of bonds broken) - Σ (bond energies of bonds formed)

It's like calculating the net cost of love: the energy spent to break up the existing relationships minus the energy gained from forming new ones.

Example: The Fiery Passion of Methane

Let's look at the combustion of methane, where methane (CH4) reacts with oxygen (O2) to form carbon dioxide (CO2) and water (H2O). This is like a fiery love triangle!

CH4(g) + 2O2(g) → CO2(g) + 2H2O(g)

To calculate the enthalpy change, we need to:

Draw the love map: Draw the Lewis structures of the molecules to see who's bonded to whom.
Identify the heartbreaks and new romances: Which bonds are broken and which are formed?
Look up the bond strengths: Find the bond energies for each bond.
Calculate the energy toll: Calculate the total energy required to break the existing bonds.
Calculate the energy payoff: Calculate the total energy released in forming new bonds.
Find the net energy change: Subtract the energy released from the energy absorbed to find ΔH.
In this case, the combustion of methane is exothermic, releasing a lot of energy (like a passionate explosion!).

Important Notes:

Bond energies are average values, so our love-o-meter isn't always perfectly accurate.
This method gives us an estimate of the enthalpy change, not an exact value.
We're assuming our atoms are in the gaseous state, like they're floating on air with love.

Case Study: Hydrogen Fuel Cells - A Sustainable Love Story

Hydrogen fuel cells are a promising technology for clean energy. They work by reacting hydrogen (H2) and oxygen (O2) to form water (H2O). This is like a sustainable love story, where the energy released can power our world.

The reaction is highly exothermic, releasing a significant amount of energy. This highlights the potential of hydrogen fuel cells as a clean and powerful energy source.

Conclusion

Understanding bond breaking, bond making, and enthalpy change is like understanding the language of love in the atomic world. By using bond energies, we can estimate the energy flow in chemical reactions and gain insights into the forces that drive these processes. This knowledge is crucial for developing new technologies and understanding the world around us.

Chemical reactions

Physical and chemical changes

The Amazing Dance of Matter: Physical and Chemical Transformations

Imagine the universe as a grand ballroom, and matter as the dancers, constantly twirling, swirling, and changing partners. These captivating moves are what we call physical and chemical changes. Let's dive into this mesmerizing dance!

Physical Changes: A Quick Change Act

In the physical realm, matter is a bit of a chameleon, changing its appearance without altering its core identity. Think of it as a dancer changing costumes – they may look different, but they're still the same dancer underneath.

Melting Ice: Picture an ice cube gracefully transforming into liquid water, like a dancer shedding a stiff outer layer to move with more fluidity. The water molecules are just loosening up, not changing their chemical makeup.

Crushing a Can: Imagine a dancer crumpling a piece of paper. The paper changes shape, but it's still paper. Similarly, crushing a can alters its form but not its fundamental composition.

Dissolving Sugar: Think of sugar dissolving in water as a dancer blending into a crowd. The sugar molecules disperse among the water molecules, but they remain sugar. This is like a magic trick where the substance seems to disappear, but it's just hiding in plain sight.

Chemical Changes: A Whole New Character

Now, imagine a dancer undergoing a complete transformation, emerging with a new costume, new moves, and a whole new persona. That's a chemical change!

Burning Wood: Envision a log in a fireplace, crackling and transforming into ash and smoke. This dramatic change is like a dancer bursting into flames, leaving behind a completely different form.

Baking a Cake: Think of the ingredients for a cake – flour, eggs, sugar – as individual dancers. When mixed and baked, they undergo a magical transformation, emerging as a delicious new creation, much like a dance troupe merging into a unified masterpiece.

Rusting Iron: Picture a shiny iron nail slowly turning into a flaky, reddish-brown substance. This is like a dancer gradually aging and changing, becoming something different over time.

Clues to Unravel the Mystery

Sometimes, it's tricky to tell if a change is physical or chemical. Here are some clues to help you decipher the dance:

New Substance? If a completely new substance with different properties appears, it's a chemical change. It's like a dancer transforming into a different being altogether.

Reversibility? Physical changes are often reversible, like a dancer changing back into their original costume. Chemical changes are usually irreversible, like a dancer who has permanently transformed.

Energy Shift? Both types of changes involve energy, but chemical changes often involve a more dramatic energy shift, like a dancer exploding with newfound energy.

The Grand Finale

Understanding the difference between physical and chemical changes is like appreciating the nuances of a complex dance. It allows us to appreciate the constant transformations happening around us, from the melting of ice to the baking of a cake. So, next time you witness a change in matter, remember the dance of matter and try to decipher its intricate steps!

Rate of reaction

Imagine a bustling dance floor, packed with people moving to the rhythm. Each person represents a molecule, and the dance floor is our reaction space. Now, for a dance-off (a reaction!) to happen, these dancers need to bump into each other – that's the essence of collision theory!

But it's not just any bump that will do. It's got to be a collision with enough energy and the right kind of groove. Think of it like this: if two dancers collide but are moving too slowly or facing the wrong way, they'll just apologize and move on. But if they collide with enough force and in the right position, sparks fly, and they start a dazzling dance-off!

Let's break down the dance moves:

(a) The Crowd (Concentration):

The more dancers on the floor, the more likely they are to bump into each other. This is like concentration – more molecules in a given space mean more chances for collisions and a faster reaction. Imagine a mosh pit versus a slow waltz – the mosh pit's going to have a lot more action!

Exception to the Rule: Sometimes, even in a crowded space, the dancers might be really good at avoiding each other (think of a well-choreographed ballet). Similarly, some reactions don't speed up just because there are more molecules around.

Real-World Example: Think of lighting a match. The higher the concentration of oxygen in the air, the faster the match will ignite and burn.

(b) The Beat (Collision Frequency):

The tempo of the music dictates how often the dancers move and collide. This is like temperature – higher temperatures mean faster-moving molecules and more frequent collisions. Think of a fast-paced salsa versus a slow tango – the salsa dancers are going to bump into each other a lot more!

State of Matter Matters: It's easier to collide on a spacious dance floor than in a packed elevator. Similarly, reactions in the gas phase happen faster because the molecules have more room to move around.

Real-World Example: Food spoils faster at room temperature because the molecules in the food and the air are moving faster and colliding more often, leading to quicker decomposition.

(c) The Energy (Kinetic Energy):

It's not just about bumping; it's about bumping with oomph! Dancers need enough energy to really make an impact. Similarly, molecules need enough kinetic energy to break their existing bonds and form new ones.

Temperature Turns Up the Heat: A hotter dance floor means more energetic dancers. Higher temperatures mean more molecules have the energy needed to react.

Real-World Example: Think of starting a fire. You need to provide enough initial energy (like striking a match) to get the reaction going.

(d) The Doorway (Activation Energy):

Imagine a special dance floor with a raised platform. Dancers need enough energy to jump onto the platform before they can join the main dance-off. This platform is like activation energy – the minimum energy needed to start a reaction.

Catalysts Open a Side Door: Now imagine a secret staircase leading to the platform. This staircase is like a catalyst – it provides an easier way to reach the platform, requiring less energy.

Real-World Example: Enzymes in our bodies are like expert dance instructors. They help reactions happen more easily by lowering the activation energy, allowing our bodies to function efficiently.

Collision Theory: Still Grooving in 2024 and Beyond:

Collision theory is like a classic dance move that never goes out of style. It's the foundation for understanding how reactions happen, and scientists are still finding new ways to use it:

Computer Simulations: Imagine choreographing a dance with incredible precision, mapping out every step and interaction. That's what scientists do with computer simulations to study collisions in detail.
Ultrafast Lasers: It's like capturing the dance-off with a high-speed camera, freezing each moment of impact. Femto- chemistry uses lasers to study reactions on incredibly short timescales.
Atmospheric Chemistry: Think of the atmosphere as a giant dance floor where molecules are constantly colliding and reacting. Collision theory helps us understand important processes like air pollution and ozone depletion.
Materials Science: Imagine designing new materials by carefully controlling how molecules collide and interact. This is crucial for developing new technologies and products.

In the End:

Collision theory is a vibrant and essential concept in chemistry. By understanding how molecules collide, we can unlock the secrets of chemical reactions and create a better future. Now, let's get back on the dance floor and see what amazing reactions we can create!

Imagine the Chemical World as a Busy City

Think of a bustling city with cars zipping around, trying to get from one place to another. These cars are like molecules in a chemical reaction, and their destination is the product they want to become. Now, some roads in this city have really steep hills – those are like reactions with high activation energy. It takes a lot of energy for the cars to climb those hills, so traffic slows down and the reaction happens slowly.

Catalysts are Like Super-Efficient Tunnels

This is where catalysts come in! They're like clever engineers who build tunnels through those hills. Suddenly, the cars can zip through the tunnels with ease, bypassing the energy-intensive climb. That's exactly what a catalyst does: it provides an easier path for the reaction, lowering the activation energy and speeding things up. And the best part? The tunnel remains intact after the cars pass through, just like the catalyst remains unchanged after the reaction.

Let's Get Hands-On: Measuring the City's Traffic Flow

Now, how do we measure how fast these reactions are happening? It's like monitoring the traffic flow in our city. We have a few clever ways to do that:

Weighing the Cars: Imagine weighing all the cars entering and leaving a particular zone. If some cars magically disappear (like in a reaction where a gas is produced), we can track the weight loss to see how fast the reaction is going. This is like the "change in mass" method.

Counting Cars Passing Through a Tunnel: We could also monitor a specific tunnel (our catalyst!) and count how many cars pass through it over time. This is similar to measuring the volume of gas produced in a reaction.

Tracking Individual Cars: Imagine attaching trackers to some cars and monitoring their speed and progress. This is like measuring the change in concentration of reactants or products over time.

The Iodine Clock: A Dramatic Traffic Light

Imagine a traffic light that suddenly changes color, signaling a dramatic change in the city's flow. That's kind of like the iodine clock reaction. It's a two-part process:

Slow Traffic: First, some cars (iodate and hydrogen sulfite ions) have a slow, leisurely drive, gradually changing into different types of cars (iodide ions).

Sudden Rush Hour: Suddenly, these new cars (iodide ions) trigger a chain reaction, causing a massive traffic jam and a sudden color change (the blue-black complex).

By timing how long it takes for this "traffic jam" to happen, we can learn a lot about the factors that speed up or slow down the reaction.

Why Does This Matter?

Understanding these "traffic patterns" in the chemical world is essential for all sorts of things. It helps us design better medicines, create new materials, and even protect our environment. It's like having a master plan for our city, ensuring everything runs smoothly and efficiently.

This is just the beginning! The world of chemistry is full of exciting discoveries and endless possibilities. Who knows what kind of "chemical cities" we'll be exploring in the future?

Imagine a bustling dance floor, packed with energetic molecules ready to mingle. That's essentially what a chemical reaction is – a lively get-together where molecules bump and grind, exchanging partners (atoms) to form new and exciting compounds. But just like any good party, there are a few things that can make the action heat up or cool down. Let's dive into the world of collision theory and explore the factors that influence these molecular meet-and-greets, drawing inspiration from the latest scientific discoveries of 2024.

(a) Concentration: The More the Merrier

Think of it like a singles mixer: the more people crammed into the room, the higher the chance of finding a compatible partner. Similarly, increasing the concentration of reactants in a solution is like packing the dance floor – more molecules in a given space mean more chances for collisions, leading to a faster reaction.

Real-world example: Ever tried dissolving an effervescent tablet in water? Crushing the tablet is like turning a slow waltz into a fast-paced salsa – the increased surface area effectively boosts the concentration of the reactant, leading to a fizzier, faster reaction.

2024 cutting-edge: Scientists are now using microfluidic devices – tiny channels where they can precisely control the concentration of reactants – to speed up reactions for applications like on-the-spot disease diagnostics and creating custom molecules on a chip.

(b) Pressure: Turning Up the Heat

Imagine those molecules as dancers in a balloon. Squeeze the balloon (increase the pressure), and suddenly everyone's bumping into each other more often. Increasing the pressure of gaseous reactants is like shrinking the dance floor – molecules get closer, collide more frequently, and the reaction rate soars.

Real-world example: The Haber-Bosch process, a crucial industrial method for producing ammonia, relies on high pressure to force nitrogen and hydrogen gas molecules to get cozy and form the desired product.
2024 breakthrough: Researchers are exploring high-pressure reactions to create materials with extraordinary properties, like metal-organic frameworks (MOFs) that can store gases, catalyze reactions, and even deliver drugs with pinpoint accuracy.

(c) Surface Area: Expanding the Dance Floor

Picture a block of wood versus a pile of sawdust. The sawdust has a much larger surface area exposed to the air, making it easier to ignite. Similarly, increasing the surface area of a solid reactant is like expanding the dance floor – more exposed particles mean more potential collision sites and a faster reaction.

Real-world example: A log burns slower than wood shavings because the shavings offer more contact points for oxygen, leading to a faster combustion reaction.
2024 nanotech magic: Nanoparticles, with their incredibly high surface area-to-volume ratios, are revolutionizing fields like catalysis, sensing, and drug delivery by dramatically enhancing reaction rates.

(d) Temperature: Raising the Roof

Crank up the music, and the dancers move with more energy, increasing the chances of a successful pairing. Increasing the temperature is like injecting the dance floor with adrenaline – molecules move faster, collide with greater force, and are more likely to overcome the "activation energy" barrier needed to react.

Real-world example: Cooking food at higher temperatures speeds up the chemical reactions that transform raw ingredients into a delicious meal.
2024 bio-innovation: Scientists are studying how temperature affects enzyme activity, those biological catalysts that drive essential processes in living organisms, to optimize biocatalytic processes for industrial applications.

(e) Catalysts: The Ultimate Matchmakers

Imagine a skilled dance instructor guiding the dancers into perfect pairings. Catalysts are like those instructors – they provide an alternative reaction pathway with a lower activation energy, making it easier for molecules to react, even at the same temperature.

Real-world example: Catalytic converters in cars use precious metal catalysts to transform harmful exhaust gases into less harmful substances, cleaning up our air.
2024 catalyst revolution: Researchers are developing new catalysts, like single-atom catalysts with maximum efficiency and biocatalysts for sustainable synthesis, to drive innovation in various fields.

Case Study: Enzyme Catalysis in Biofuel Production

Biofuels offer a promising alternative to fossil fuels, and enzymes play a crucial role in their production. Imagine these enzymes as expert choreographers, breaking down complex plant materials into simpler sugars that can be fermented into bioethanol. Scientists are constantly tweaking these choreographers, optimizing their performance to make biofuel production more efficient and sustainable.

The Grand Finale

Understanding how these factors influence reaction rates is like having the keys to the molecular disco. Whether you're a chemist, a materials scientist, a biologist, or simply curious about the world around you, collision theory provides a fascinating glimpse into the intricate dance of molecules that drives countless processes in our universe.

Reversible reactions and equilibrium

Reversible Reactions: A Chemical Tango

Imagine a dance floor where couples constantly form and break apart, swaying back and forth in a delicate balance. This is the essence of a reversible reaction – a chemical tango where reactants and products continuously transform into each other.

The Dance of Molecules

In the world of chemistry, some reactions are like one-way streets, proceeding relentlessly towards the formation of products. But reversible reactions are more like a two-way street, allowing molecules to switch partners and reverse their steps.

We represent this chemical dance with a special symbol: $\rightleftharpoons$. Think of it as a double arrow, indicating that the reaction can proceed in both directions.

$$A + B \rightleftharpoons C + D$$

In this equation, A and B can waltz together to form C and D, but C and D can also reunite to recreate A and B. It's a continuous cycle of formation and reformation.

Factors Influencing the Dance

Just like a dance floor can be influenced by external factors, the direction of a reversible reaction can be swayed by changes in conditions:

Temperature: Turning up the heat can favor one direction of the dance, while cooling things down might favor the opposite direction.
Pressure: For reactions involving gases, increasing the pressure is like shrinking the dance floor, forcing the molecules closer together and potentially favoring the side with fewer molecules.
Concentration: Adding more of a reactant or product is like flooding the dance floor with one type of partner, influencing the equilibrium of the dance.

Hydrated and Anhydrous Compounds: A Colorful Example

Let's dive into a specific example to see how these factors play out in the fascinating world of hydrated and anhydrous compounds.

Hydrated compounds are like crystals that have trapped water molecules within their structure. These water molecules are called "water of crystallization." When heated, these compounds release their water molecules and become anhydrous compounds, like dancers shedding their costumes.

Hydrated compound ⇌ Anhydrous compound + Water

This is a reversible reaction, meaning the anhydrous compound can regain its water molecules and transform back into its hydrated form.

Le Chatelier's Principle: The Dance Floor Manager

To understand how changes in conditions affect this dance, we need to introduce Le Chatelier's principle. This principal acts like a dance floor manager, ensuring that the dance maintains a delicate balance. It states that if a change of condition is applied to a system in equilibrium, the system will shift in a direction that relieves the stress.

Heat and the Copper (II) Sulfate Tango

Let's take hydrated copper (II) sulfate, a beautiful blue crystal, as our dance partner. When heated, it loses its water molecules and transforms into anhydrous copper (II) sulfate, a white powder.

$CuSO4.5H2O(s) \rightleftharpoons CuSO4(s) + 5H2O(g)$

Heating this compound is like turning up the heat on the dance floor. According to Le Chatelier's principle, the system will shift towards the direction that absorbs the heat, which is the forward reaction (dehydration). This is because the dehydration reaction is endothermic – it absorbs heat to proceed.

Cooling the system, on the other hand, will favor the reverse reaction (hydration), as it releases heat (exothermic).

Adding Water: A Splash of Change

Adding water to anhydrous copper (II) sulfate is like introducing a new group of dancers to the floor. The system will shift towards the direction that consumes the water, which is the forward reaction (hydration).

Cobalt (II) Chloride: A Humidity Sensor

Anhydrous cobalt (II) chloride is another fascinating dancer. It's blue when dry but turns pink when exposed to moisture, making it a natural humidity indicator.

$$CoCl_2(s) + 6H_2O(l) \rightleftharpoons CoCl_2.6H_2O(s)$$

In dry conditions, the cobalt (II) chloride remains blue. As humidity increases, it eagerly absorbs water molecules from the air and transforms into its pink hydrated form.

Real-World Applications: The Dance of Life

Reversible reactions are not just confined to the laboratory; they play a vital role in our everyday lives:

Desiccants: Anhydrous compounds that readily absorb water are used as desiccants to keep things dry. Think of those little silica gel packets found in shoe boxes or electronics packaging.
Plaster of Paris: This type of anhydrous calcium sulfate hydrates to form gypsum, a harder compound used in making casts and molds.
Gas Masks: Some gas masks utilize hydrated copper (II) sulfate to absorb water vapor from the wearer's breath, preventing moisture buildup inside the mask.

Beyond the Basics

This is just the beginning of our exploration of reversible reactions. There's a whole world of fascinating concepts to discover, including dynamic equilibrium, enthalpy and entropy changes, Gibbs free energy, and the effect of pressure.

Conclusion

Reversible reactions are a fundamental concept in chemistry, a continuous dance of molecules transforming between reactants and products. By understanding the factors that influence this dance, we can control and manipulate chemical reactions for various applications, from industrial processes to everyday life. So, let's continue to explore this fascinating world of chemical tango and unravel the secrets of reversible reactions.

Reversible Reactions and Equilibrium: A Chemical Tango

Imagine a dance floor where molecules are the dancers. Some couples pair up (reactants forming products), while others break apart (products reverting to reactants). This constant back-and-forth is the essence of a reversible reaction. Now, picture the music reaching a crescendo, the dancers finding a rhythm, and the number of couples forming and breaking apart becoming equal. This is equilibrium – a dynamic state of balance where the reaction seems to have paused, but the dance continues.

1. Equilibrium: Finding the Balance in a Chemical Dance

Equilibrium in a reversible reaction is like a perfectly choreographed dance where two things happen:

The Forward and Reverse Steps Match: Imagine dancers moving seamlessly between forming pairs and breaking apart at the exact same rate. This is what happens at equilibrium – the forward and reverse reactions occur at equal speeds.
Concentrations Remain Steady: Even though the dance continues, the number of "couples" (products) and "singles" (reactants) on the dance floor stays constant. This means the concentrations of reactants and products remain unchanged over time.

2. The Haber Process: An Industrial Waltz

The Haber process, used to produce ammonia for fertilizers, is a prime example of this chemical dance. Nitrogen and hydrogen molecules waltz together to form ammonia, but ammonia can also break back down into its original partners:

$N_2 (g) + 3H_2 (g) \rightleftharpoons 2NH_3 (g)$

Think of it like a dance with a revolving door – some molecules are entering the "ammonia room," while others are exiting. At equilibrium, the number of molecules entering and exiting the room is the same, keeping the concentration of ammonia stable.

3. Factors Affecting Equilibrium: Changing the Music

Just like different music can change the energy and flow of a dance, several factors can influence the position of equilibrium:

Concentration: Adding more reactants is like adding more dancers – it encourages more couples to form (shifting equilibrium towards products).
Pressure: Imagine the dance floor shrinking. Couples (taking up less space) become favored over singles, shifting the equilibrium towards the side with fewer gas molecules.
Temperature: Turning up the heat can either energize the dancers to form more couples (endothermic reaction) or cause them to break apart to cool down (exothermic reaction).
Catalyst: A catalyst is like a skilled dance instructor, speeding up both the pairing and breaking apart of couples without affecting the final number of each.

4. Le Chatelier's Principle: Maintaining the Rhythm

Le Chatelier's principle is the "dance etiquette" of equilibrium. It states that if you disrupt the equilibrium (change the music, floor size, or temperature), the system will adjust to restore the balance.

5. Equilibrium Constant (K): The Dance Scorecard

The equilibrium constant (K) is like a scorecard for the dance, telling you the ratio of products to reactants at equilibrium. A high K value means more "couples" (products) are present, while a low K value indicates more "singles" (reactants).

6. Applications of Equilibrium: Beyond the Dance Floor

The concept of equilibrium extends far beyond chemistry labs:

Industrial Chemistry: Optimizing production of chemicals like ammonia.
Environmental Science: Understanding acid rain formation and gas solubility in water.
Biological Systems: Enzyme activity and oxygen transport in the blood.
Medicine: Drug interactions and electrolyte balance in the body.

7. Advanced Concepts: The Choreography of Complex Dances

Beyond the basics, there are more intricate steps in this chemical dance:

Reaction Quotient (Q): A snapshot of the dance floor at any moment, indicating the direction the reaction needs to move to reach equilibrium.
Solubility Equilibrium: The dance of dissolving salts in water.
Acid-Base Equilibrium: The interplay between acids and bases, like a dance between partners with different strengths.

8. Conclusion: The Ever-Moving Dance of Chemistry

Reversible reactions and equilibrium reveal the dynamic nature of the chemical world. It's a constant dance of forming and breaking bonds, shifting and adjusting to maintain balance. Understanding this dance allows us to control and predict chemical reactions, impacting fields from medicine to environmental science.

Imagine a Chemical Dance Floor

Picture a bustling dance floor at a club. On one side, you have couples forming (reactants becoming products). On the other side, couples are breaking up (products reverting to reactants). This is a reversible reaction, a constant back and forth. Now, imagine different things happening at this club:

(a) Turning Up the Heat (Temperature)

Endothermic Reaction (Heat is a "reactant"): Imagine the dance floor getting chilly. To warm up, more people pair up and dance vigorously (more products form) to generate body heat. If the AC kicks in and it gets too cold, couples start breaking up to conserve energy (more reactants form).

Exothermic Reaction (Heat is a "product"): Now the dance floor is getting too hot! Couples start breaking up to cool down (more reactants form). If the club cools down, more people pair up to dance and generate some warmth (more products form).

Think of the Haber process (making ammonia) like a packed dance floor generating a lot of heat. Turning up the heat makes people uncomfortable, so they break up (less ammonia).

(b) Feeling the Pressure (Pressure)

Imagine the dance floor getting crowded. To make more space, people start forming groups (shifting to the side with fewer molecules). If the crowd thins out, people break away from their groups and dance individually (shifting to the side with more molecules).

In the Haber process, increasing the pressure is like making the dance floor more crowded. People pair up to save space (more ammonia).

(c) New Arrivals and Departures (Concentration)

More reactants: A new group enters the club, eager to dance. This encourages more couples to form (more products).

More products: The dance floor is filled with couples. Some people get tired and leave, encouraging others to break up and take a break (more reactants).

In the Haber process, adding more nitrogen or hydrogen is like bringing in a fresh batch of dancers. More couples form (more ammonia).

(d) The DJ (Catalyst)

The DJ sets the mood and keeps the energy going, but they don't force anyone to dance or break up. They just help things move along.

In the Haber process, the iron catalyst is like the DJ. It speeds up the reaction but doesn't change how many couples are on the dance floor (doesn't affect the amount of ammonia).

Where Do the Dancers Come From? (Haber Process Sources)

(a) Hydrogen: The Natural Gas Pipeline

Imagine a pipeline supplying the club with dancers (hydrogen). This pipeline is fed by a massive natural gas field (methane). The methane is transformed into dancers (hydrogen) through a two-step process:

Steam Methane Reforming: Methane molecules are like groups of friends entering the club. They mingle with steam (water vapor) in a heated room with a nickel catalyst (the bouncer). This breaks them up into individual dancers (hydrogen) and some carbon monoxide.

Water-Gas Shift Reaction: The carbon monoxide is like a wallflower. It's encouraged to mingle with more steam to form carbon dioxide (leaves the club) and even more dancers (hydrogen).

(b) Nitrogen: Straight from the Air

Imagine a giant air vent pumping fresh air into the club. This air is like a mix of different people, but we only want the nitrogen dancers. So, we use a special machine:

Air Liquefaction: The air is cooled down until it becomes a liquid, like a crowded room where everyone is squeezed together.

Fractional Distillation: This machine is like a sorting hat. It separates the nitrogen dancers from the oxygen dancers based on their "dance moves" (boiling points). Nitrogen is more volatile (boils off first), so it's collected for the dance floor.

Why these sources?

Abundance and Cost: Natural gas and air are plentiful and cheap, like having a huge supply of eager dancers and a free venue.

Convenience: We've been using these sources for a long time, so we have the equipment and know-how to get the dancers we need efficiently.

This whimsical analogy explains the complex concepts of equilibrium shifts and the Haber process in a fun and memorable way. Remember, the specific conditions and choices of sources can vary depending on the "club" (location) and its specific needs.

The Haber Process: A Love Story of Nitrogen and Hydrogen (and a Little Bit of Iron)

Imagine a bustling singles bar, but instead of humans, it's filled with nitrogen and hydrogen atoms. They're bumping into each other, flirting, but nothing serious is happening. They need a little push, a little something to spark a real connection. Enter: the Haber Process, the ultimate wingman (or wing-process, if you will).

This process isn't about forcing a relationship – it's about setting the mood. First, it cranks up the heat to a cozy 450°C (723 K) – not too hot, not too cold, just right for these shy atoms. Then, it turns up the pressure to a whopping 20,000 kPa (200 atm), getting everyone nice and close. Finally, it introduces the smooth operator, iron (with a dash of potassium oxide and aluminum oxide for extra charm).

Iron acts like the perfect chaperone, bringing nitrogen and hydrogen together in a way they could never manage on their own. Sparks fly, bonds form, and voila! We have ammonia (NH₃), the lovechild of this atomic romance.

But like any good relationship, it's a two-way street. Ammonia can break up too (the reverse reaction), so the Haber Process has to keep things balanced. It's a delicate dance of temperature, pressure, and catalytic matchmaking.

Why is this love story so important?

Well, ammonia is a VIP in the world. It's the key ingredient in fertilizers that help grow the food we eat. It's used in explosives, various chemicals, and even as a refrigerant. Basically, ammonia keeps our modern world running.

But there's a catch...

This love story has a bit of a dark side. The Haber Process is a bit of an energy hog, and it can lead to pollution. It's like a romantic getaway that leaves a hefty carbon footprint.

The future of this love triangle?

Scientists are working on making this relationship more sustainable. They're looking for ways to use renewable energy to power the process and finding new catalysts that work their magic at lower temperatures and pressures. It's all about finding a way for nitrogen and hydrogen to find love in a way that's good for the planet.

Want to learn more about other chemical matchmakers? Stay tuned for the story of the Contact Process and how it helps create sulfuric acid!

Sulfuric Acid: The Unsung Hero of Chemistry (and How We Make It)

Forget superheroes with capes and flashy powers. The real champion of the chemical world is a humble, colorless liquid – sulfuric acid. This stuff is everywhere, quietly powering industries from fertilizers that feed us to the metals that build our world.

But where does this powerhouse come from? It's not mined or harvested; it's created through a process so ingenious it deserves a standing ovation. Enter the Contact Process, a chemical ballet in three acts where sulfur and oxygen tango to create the king of acids.

Act 1: Summoning the Sulfur Dioxide

Imagine a fiery underworld where sulfur, mined from the earth or snatched from the fumes of oil refineries, is set ablaze. This isn't some haphazard bonfire; it's a controlled burn where sulfur meets oxygen in a passionate embrace, giving birth to sulfur dioxide (SO_2).

But there's another way to summon SO_2. Picture massive ovens where sulfide ores like pyrite – fool's gold – are roasted until they surrender their sulfurous treasure. It's like medieval alchemy, transforming rock into a vital ingredient.

Act 2: A Catalytic Romance

Now, the heart of the drama: transforming SO_2 into sulfur trioxide (SO_3). This is where the "contact" happens, a carefully choreographed encounter between SO_2, oxygen, and a special matchmaker – the catalyst.

Imagine a dance floor heated to 450°C (that's over 840°F!). SO_2 molecules, energized by the heat, are eager to partner with oxygen, but they need a little nudge. Enter vanadium pentoxide (V_2O_5), the smooth operator who facilitates the connection. This catalyst lowers the energy barrier, making the reaction happen faster and with more finesse.

But there's a twist! This dance is a delicate balance. Too much heat, and the SO_3 molecules get overwhelmed and break apart. Too little pressure, and they can't get close enough to connect. It's like finding the sweet spot in a salsa – not too fast, not too slow, just right.

Act 3: The Grand Finale

In the final act, SO_3 takes a dip in concentrated sulfuric acid, creating a super-concentrated form called oleum. Then, with a careful addition of water, the grand transformation is complete: sulfuric acid (H_2SO_4) emerges, ready to take its place as the workhorse of industry.

Why This Matters

The Contact Process isn't just a chemistry lesson; it's a testament to human ingenuity. By understanding the delicate interplay of temperature, pressure, and catalysts, we've harnessed the power to create a substance that touches nearly every aspect of our lives. So next time you see a car, a building, or even a bag of fertilizer, remember the unsung hero and its remarkable journey from sulfur and air to the king of chemicals.

Imagine a Factory Run by Tiny Chefs

Forget the cold, hard facts for a moment. Imagine a bustling factory filled with tiny chefs, all clad in white coats and ridiculously tall hats. Their mission? To cook up one of the most important ingredients in the world: sulfuric acid.

But these aren't your average chefs. They're masters of manipulating molecules, using a recipe passed down through generations – the Contact Process.

The Star of the Show: Sulfur Dioxide

Our star ingredient, sulfur dioxide (SO2), arrives at the factory as a bit of a hothead – a pungent gas with a bit of a temper. The chefs' challenge is to transform this fiery character into something more useful: sulfur trioxide (SO3).

The Secret Ingredient: Vanadium(V) Oxide

Now, this transformation isn't easy. It's like trying to convince a grumpy cat to do a tap dance. That's where the secret ingredient comes in: vanadium(V) oxide (V2O5). Think of it as the magical spice that makes the whole process work.

This special spice acts as a catalyst, a kind of molecular matchmaker. It persuades the sulfur dioxide and oxygen (O2) molecules to come together in a cozy dance, forming sulfur trioxide.

A Delicate Balancing Act

The chefs have to be careful, though. If the factory gets too hot, the sulfur trioxide starts to break down, undoing all their hard work. But if it's too cold, the whole process slows to a crawl. They need to find the perfect temperature – a Goldilocks zone where the reaction hums along nicely.

Pressure also plays a role. A bit of pressure helps nudge the molecules in the right direction, but too much, and the whole factory risks exploding!

The Grand Finale

Once the sulfur trioxide is formed, it's time for the final act. The chefs carefully combine it with water, creating the star of the show: sulfuric acid! This versatile liquid is then shipped off to other factories, where it's used to make everything from fertilizers and detergents to batteries and plastics.

A Recipe for Success

The Contact Process is a testament to human ingenuity. By understanding the delicate dance of molecules and the power of catalysts, we've learned to create essential chemicals that improve our lives in countless ways. And who knows? Maybe those tiny chefs are still hard at work, cooking up a batch of sulfuric acid right now!

Redox Reactions: The Chemical Dance of Electron Swapping

Imagine a bustling dance floor where atoms are constantly swapping partners (electrons!). That's essentially what's happening in redox reactions, these fundamental processes that power everything from the slow waltz of rust formation to the energetic jive of cellular energy production. Let's break it down with a touch of flair and a dash of imagination.

1. Oxidation Numbers: Atoms with Identity Crises

Think of oxidation numbers as atomic "aliases" – they help us track electrons as they move between atoms during their chemical dance-off. These numbers represent the charge an atom would have if it were a bit of a drama queen and hogged all the electrons in its bonds.

Rules for Unmasking these Atomic Aliases:

Free Elements: Atoms flying solo have an oxidation number of zero. They're content being themselves, like a noble gas chilling in its own space. (Think O in O_2, or Fe just hanging out as metallic iron.)
Monatomic Ions: These are atoms that have already gained or lost electrons and are now sporting a full-blown charge. Their oxidation number is simply their charge. (Na^+ is +1, like a proud cheerleader with an extra pom-pom, while Cl^- is -1, like a grumpy librarian shushing everyone.)
Hydrogen: Usually +1, the eager beaver, except when it's hanging out with metals in metal hydrides, where it becomes -1, the rebellious teenager.
Oxygen: Typically, -2, the queen bee, except in peroxides (-1, sharing the crown) and super- oxides (-½, in a power-sharing agreement).
Group 1 and 2 Metals: Always +1 and +2, the reliable members of the periodic table, always showing up with the same energy.
Halogens: Usually -1, the sneaky ninjas, except when they're with oxygen or a more electronegative halogen, in which case they might switch things up.
The Balancing Act: In a neutral compound, the sum of all the oxidation numbers must be zero. It's like a perfectly balanced seesaw.

In a polyatomic ion, the sum equals the ion's charge – a bit like a team effort where everyone contributes to the overall score.
Examples:

H_2O: H (+I), O (-II) – a classic partnership, like peanut butter and jelly.
CO_2: C (+IV), O (-II) – carbon showing off its versatility, like a chameleon changing colors.
$KMnO_4$: K (+I), Mn (+VII), O (-II) – a complex dance troupe with manganese as the star performer.

Roman Numerals: The Fancy Titles

We use Roman numerals to give elements with multiple personalities (like those tricky transition metals) their proper titles. Think of it as giving them fancy nicknames to distinguish their different roles.

$FeCl_2$: Iron (II) chloride (Fe^{2+}) – Iron in its "chill" mode.
$FeCl_3$: Iron (III) chloride (Fe^{3+}) – Iron feeling a bit feistier.

2. Redox Reactions: The Electron Exchange Party

Redox reactions are like a grand ball where atoms come together to exchange electrons and change their oxidation numbers. It's a constant give-and-take, a swirling dance of electron transfer.

Key Players in this Atomic Ballroom:

Oxidation: Losing electrons, like shedding an old costume to reveal a more glamorous outfit. The oxidation number increases.
Reduction: Gaining electrons, like acquiring a shiny new accessory. The oxidation number decreases.
Reducing Agent: The generous giver of electrons, causing another atom to be reduced. It sacrifices itself for the sake of another's transformation.
Oxidizing Agent: The electron grabber, causing another atom to be oxidized. It's a bit of a diva, always wanting more.

Mnemonic Devices: Remember the Dance Moves!

OIL RIG: Oxidation Is Loss (of electrons), Reduction Is Gain (of electrons).
LEO the lion says GER: Lose Electrons Oxidation, Gain Electrons Reduction.

Example:

Picture zinc and copper (II) sulfate meeting on the dance floor:

$Zn(s) + Cu\ SO_4(AQ) \rightarrow Zn\ SO_4(AQ) + Cu(s)$

Zn: Loses two electrons (oxidation), its oxidation number goes from 0 to +II. It's the life of the party, donating electrons and transforming into Zn^{2+}. (Reducing agent)
Cu^{2+}: Gains two electrons (reduction), its oxidation number drops from +II to 0. It's the wallflower who gets a makeover, becoming Cu. (Oxidizing agent)

3. Redox Reactions: Beyond the Oxygen Obsession

Historically, redox reactions were all about oxygen – gaining it was oxidation, losing it was reduction. But now we know it's really about those electrons! Oxygen is often involved, but it's not the whole story.

Connecting the Dots:

Think of oxygen as a popular dance partner. When it joins the dance, it often brings along its entourage of electrons. So, gaining oxygen often means gaining electrons (reduction), and losing oxygen often means losing electrons (oxidation).

4. Types of Redox Reactions: Different Dance Styles

Redox reactions come in various forms, each with its own unique choreography:

Combination Reactions: Two or more substances come together to form a single product, like a group dance where everyone merges into a synchronized routine.
Decomposition Reactions: A single compound breaks down into simpler substances, like a dance crew splitting off into solo performances.
Displacement Reactions: One element kick another out of a compound, like a dance battle where the winner steals the spotlight.
Combustion Reactions: A substance reacts with oxygen (usually), releasing heat and light, like a fiery tango that sets the dance floor ablaze.
Disproportionation Reactions: A single species undergoes both oxidation and reduction, like a dancer who can seamlessly switch between leading and following.

5. Applications of Redox Reactions: Redox in the Real World

Redox reactions aren't just confined to the chemistry lab; they're all around us!

Biological Processes: Respiration, photosynthesis – these are the intricate dances of life, fueled by redox reactions.
Energy Production: Combustion engines, batteries, fuel cells – they all rely on the energy released from redox reactions.
Metallurgy: Extracting metals from ores often involves redox processes, like a treasure hunt where you have to unlock the metal from its rocky prison.
Corrosion: Rusting – the slow, destructive dance of iron with oxygen and water.
Environmental Chemistry: Redox reactions play a key role in cleaning up pollutants and maintaining the delicate balance of our environment.

6. Balancing Redox Reactions: Keeping the Dance Floor Orderly

Balancing redox reactions can be a bit like choreographing a complex dance routine. You need to make sure everyone has a partner and the steps are all in sync. The half-reaction method is a common technique, where you break down the reaction into two parts (oxidation and reduction), balance each separately, and then combine them, making sure the electron transfer is balanced.

7. Case Studies: Redox in Action

Rusting of Iron: A classic redox tale of iron's slow demise in the presence of oxygen and water.
Photosynthesis: The amazing dance of plants using sunlight to convert carbon dioxide and water into sugar and oxygen, a redox process that sustains life on Earth.
By understanding the intricate steps of redox reactions, we can appreciate the elegant choreography of the chemical world around us.

Oxidation and Reduction: It's Like Life, But with Electrons (2024 Update)

Imagine electrons as tiny, energetic gossips, always eager to jump from one atom to another. Oxidation and reduction are basically the stories of these electron escapades. They're like two sides of the same coin, a dynamic duo that's always up to something in the world of chemistry. Whether it's a fire crackling, metal rusting, or a plant soaking up the sun, redox reactions are the secret puppet masters behind the scenes.

1. Oxidation: Losing those Precious Electrons

Think of oxidation as an atom having a bit of an existential crisis. It's like losing your phone, your keys, and your wallet all at the same time – a loss of precious electrons!

The Oxygen Addict: In the old days, chemists thought oxidation was all about oxygen glomming onto other elements. Like when iron gets rusty, its basically oxygen clinging on for dear life, turning the iron into a flaky mess.

Electron Escape Artists: But then, chemists got wiser and realized oxidation is really about electrons making a run for it. It's like magnesium, a generous soul, giving away its electrons to chlorine in a dramatic ionic exchange.

The Oxidation Number Game: Now, imagine each atom has a secret code, its oxidation number. When this number goes up, it means the atom has lost some of its electron buddies. It's like going into debt – the higher the number, the more you owe (in this case, electrons!).

2. Reduction: The Electron Magnet

Reduction is the opposite of oxidation. It's like finding a treasure chest full of shiny new electrons!

Oxygen's Farewell: Sometimes, reduction is about saying goodbye to oxygen. Imagine copper oxide, feeling a bit suffocated, letting go of oxygen and becoming pure, shiny copper again.

Electron Welcome Party: But more often, reduction is about atoms happily accepting those runaway electrons. It's like chlorine throwing a party and all the electrons rushing in to join the fun.

The Oxidation Number Plunge: Remember that secret code, the oxidation number? In reduction, this number goes down, meaning the atom is gaining electrons. It's like paying off your debt and feeling a sense of relief.

Redox Reactions: The Big Picture

Two-for-One Deal: Oxidation and reduction always happen together. It's like a dance, where one partner loses a step (electrons) and the other gains it.

Electron Express: Redox reactions are all about electrons hopping from one place to another, like tiny couriers delivering messages.

The Push and Pull: The oxidizing agent are the electron thief, causing oxidation. The reducing agent is the electron donor, causing reduction. They're like the yin and yang of the electron world.

Redox in Action: Real-Life Drama

Burning Desire (Methane Combustion): When methane burns, it's like a wild party where carbon loses its electrons (oxidation) and oxygen grabs them (reduction). The result? Carbon dioxide, water, and a whole lot of heat!

Rusty Relationships (Iron Corrosion): Iron and oxygen have a complicated relationship. Oxygen steals electrons from iron, leaving it rusty and heartbroken.

Plants: The Electron Wizards (Photosynthesis): Plants are like tiny solar panels, using sunlight to power amazing redox reactions. They take carbon dioxide and water, shuffle some electrons around, and create sugary treats (glucose) and oxygen.

Breathing Easy (Cellular Respiration): We breathe in oxygen, which helps us break down glucose and release energy. It's like a reverse photosynthesis party, where oxygen grabs electrons and carbon lose them.

Batteries: Electron Storage Units: Batteries are like electron banks, storing them up and releasing them when we need power. Redox reactions are the key to this energy magic.

The Future is Redox!

From clean energy to life-saving medicines, redox reactions are shaping our future. By understanding this electron dance, we can create new technologies and solve some of the world's biggest challenges. So, let's raise a glass to oxidation and reduction – the tiny but mighty forces that keep our world spinning!

Redox Reactions: The Electron Tango

Imagine a bustling dance floor where electrons are the stars, twirling and swirling from one partner to another. That's essentially what happens in a redox reaction – a captivating chemical dance where electrons are transferred between atoms. One partner loses an electron (oxidation) while the other gains one (reduction). It's a bit like a captivating game of chemical "tag" where electrons are the coveted prize.

Oxidation Numbers: The Electron Scorecard

To keep track of this electron exchange, we use oxidation numbers – a clever system for assigning "scores" to atoms in a compound. These numbers help us see which atoms are gaining and losing electrons, revealing the intricate steps of the redox dance.

Rules of the Game:

Free Elements: Atoms flying solo have an oxidation number of zero. They're like independent dancers waiting to join the party. (Think O2, Cl2, Fe)
Monatomic Ions: These ions have a score equal to their charge. They're like dancers with a clear identity, either positive or negative. (Think Na+, Cl-)
Neutral Compounds: In a neutral compound, the oxidation numbers add up to zero. It's like a balanced dance floor with equal numbers of positive and negative dancers. (Think H2O)
Polyatomic Ions: These ions have a combined score equal to their charge. They're like dance troupes with a collective identity. (Think SO42-)
Special Elements: Some elements have signature moves:
Group 1 Metals (Alkali Metals): Always +1, like enthusiastic dancers leading the charge.
Group 2 Metals (Alkaline Earth Metals): Always +2, a bit more grounded but still eager to participate.
Hydrogen: Usually +1, a versatile dancer, but sometimes -1 when paired with a metal.
Oxygen: Usually -2, a graceful and reliable partner, but occasionally -1 or even positive when dancing with fluorine.
Fluorine: Always -1, a strong and assertive dancer who always takes the lead.
Halogens (Cl, Br, I): Usually -1, but can change their tune when paired with oxygen or a more electronegative halogen.

Spotting the Redox Dance:

Assign Scores: Give each atom its oxidation number.
Compare Scores: Look for changes in oxidation numbers between reactants and products.
Identify the Moves: If an atom's number goes up, it's been oxidized (lost electrons). If it goes down, it's been reduced (gained electrons).
Confirm the Dance: If there are changes in oxidation numbers, you've got yourself a redox reaction!

Real-Life Redox Stories:

Rusty Romance: When iron rusts, it's a slow dance with oxygen, resulting in a reddish-brown iron oxide (rust). Iron loses electrons (oxidized) while oxygen gains them (reduced).
Cellular Respiration: Our bodies are powered by redox reactions! In cellular respiration, glucose is oxidized, releasing energy, while oxygen is reduced. It's a vital dance that keeps us alive.

The Grand Finale

Understanding redox reactions is like appreciating the intricate choreography of the chemical world. By mastering oxidation numbers, we can decode these electron dances and gain a deeper understanding of the fascinating processes that shape our world.

Redox Reactions: Where Electrons Dance and Colors Change

Imagine a bustling dance floor where electrons are the life of the party, constantly switching partners and creating a whirlwind of energy. That's kind of what a redox reaction is like! "Redox" is just shorthand for "reduction-oxidation," which basically means electrons are being passed around like delicious appetizers at a fancy gathering.

These electron exchanges aren't just some random chemical shuffle – they're the driving force behind everything from the way our bodies turn food into energy to how fireworks paint the night sky with dazzling colors.

Potassium Manganate (VII) and Potassium Iodide: The Colorful Detectives

Now, let's meet our star detectives in the world of redox reactions: acidified potassium manganate (VII) ($KMnO_4$) and aqueous potassium iodide (KI). These two chemicals are like chameleons, changing their colors to reveal the secrets of electron transfer.

$KMnO_4$: The Purple Queen of Oxidation

Imagine $KMnO_4$ as a regal queen, draped in a majestic purple robe. This queen is a bit of an electron hoarder – an "oxidizing agent." She loves to snatch electrons from other molecules, and when she does, her vibrant purple fades away as if by magic. It's like she's shedding her royal attire to blend in with the crowd.

For instance, when $KMnO_4$ meets iron (II) sulfate ($FeSO_4$), it's like a grand ball where the queen steals the show. She grabs electrons from $FeSO_4$, transforming it while losing her own purple color in the process.

KI: The Unassuming Electron Donor

KI, on the other hand, is like the unassuming guest at the party who generously offers electrons to anyone in need. As a "reducing agent," KI gives away its electrons, causing a dramatic transformation. The solution turns a warm yellow-brown, like a shy wallflower suddenly stepping into the spotlight.

Think of KI reacting with hydrogen peroxide (H_2O_2) like a superhero origin story. KI swoops in to donate electrons, neutralizing the potentially harmful H_2O_2 and changing color in the process – a true sign of its heroic deed.

Oxidation and Reduction: The Dynamic Duo

Remember that electron dance floor? Well, "oxidation" is like losing your dance partner (losing electrons), while "reduction" is like gaining a new one (gaining electrons). They always go hand-in-hand, like two sides of the same coin.

Case Studies: Redox in Action

Redox reactions aren't just confined to the lab – they're happening all around us!

Rusting Iron: That rusty old bike in your garage? It's a victim of a redox reaction where iron (Fe) loses electrons to oxygen (O_2), forming iron oxide (rust). It's like iron is sharing its electrons with oxygen a bit too generously.
Burning Methane: When you light a gas stove, you're initiating a redox reaction between methane (CH_4) and oxygen (O_2). Methane loses electrons, while oxygen gains them, releasing energy in the form of heat and light. It's a fiery dance of electron exchange!
Photosynthesis: Plants are masters of redox reactions! In photosynthesis, they use sunlight to power a redox reaction where carbon dioxide (CO_2) gains electrons and water (H_2O) loses them, creating glucose (sugar) and oxygen. It's like plants are playing matchmaker for electrons, and we get to breathe the oxygen they produce!

The Grand Finale

Redox reactions are the silent, unseen forces shaping our world. They're the reason we can breathe, the reason we have energy, and the reason our world is so full of vibrant colors. By understanding how electrons move and change partners in this intricate dance, we can unlock the secrets of chemistry and appreciate the magic happening all around us.

Acids, bases and salts

The characteristic properties of acids and bases

Acids, Bases, and Alkalis: A Colorful Symphony of Chemistry

Imagine the world of chemistry as a grand orchestra, filled with diverse instruments each playing their unique part. Among these, acids, bases, and alkalis form a vibrant trio, their interactions creating a symphony of reactions that drive countless processes in nature and industry.

1. The Aqueous Dance of Acids and Alkalis

Picture a bustling dance floor where molecules twirl and interact. When acids join this dance in water, they release tiny, energetic hydrogen ions (H^+), like sparks of energy. These ions give acids their signature sour taste, and they can even engage in lively exchanges with certain metals. The more hydrogen ions present, the more acidic the solution becomes, like turning up the volume on the sourness.

On the other side of the dance floor, alkalis gracefully release hydroxide ions (OH^-) as they move. These ions bring a bitter taste and a slippery feel to the party. The more hydroxide ions present, the more alkaline the solution, like adding a smooth bass line to the chemical music.

Examples:

Acids:

Hydrochloric acid (HCl): The tangy conductor of digestion in our stomachs.
Sulfuric acid (H_2SO_4): The powerhouse found in car batteries, driving energy forward.
Acetic acid (CH_3COOH): The subtle note of vinegar, adding zest to our salads.

Alkalis:

Sodium hydroxide ($NaOH$): The vigorous cleaner, unclogging drains with its power.
Potassium hydroxide (KOH): The gentle soap-maker, creating a soothing lather.
Calcium hydroxide ($Ca(OH)_2$): The sturdy builder, forming the backbone of mortar and plaster.

2. The Proton Exchange: A Deeper Connection

In 1923, two brilliant chemists, Johannes Nicolaus Brønsted and Thomas Martin Lowry, unveiled a deeper understanding of this chemical dance. They proposed that acids and bases interact through a delicate exchange of protons (H^+ ions), like dancers passing a flower.

Acids became known as proton donors, generously offering a proton to another molecule. In doing so, they transform into their conjugate base, like a dancer changing partners.

Bases, on the other hand, became proton acceptors, gracefully receiving a proton from another molecule. This transforms them into their conjugate acid, like a dancer finding a new partner.

This elegant theory expanded our understanding beyond the watery dance floor, allowing us to explore acid-base interactions in diverse environments.

3. Bases and Alkalis: A Family Affair

Bases, a broad family of compounds, are the peacemakers, always ready to neutralize the fiery acids. They often appear as oxides or hydroxides of metals.

Alkalis are a special branch of this family, those bases that readily dissolve in water. They are typically formed from the lively alkali metals (Group 1) or the more grounded alkaline earth metals (Group 2).

Properties of Bases:

Neutralizers: They calm the acidic storm, forming salts and water in the process.
High pH: Their pH values rise above 7, a testament to their alkaline nature.
Bitter taste: A sharp contrast to the sourness of acids.
Slippery feel: A smooth touch, like soap between your fingers.
Electrical conductors: Their aqueous solutions can carry an electrical current, thanks to the presence of ions.

Alkalis share all these properties, but with an added zest:

High reactivity: They are eager to interact, sometimes causing burns if handled carelessly.
Strong base formation: Group 1 hydroxides create powerful alkalis, while Group 2 hydroxides are a bit milder.

4. Real-World Rhythms: Case Studies and Applications

The symphony of acids, bases, and alkalis plays out in countless scenarios around us:

Acid Rain: A somber tune where sulfur dioxide and nitrogen oxides, released into the atmosphere, transform into acidic rain, harming forests, lakes, and even buildings.
Antacids: A soothing melody where basic compounds like calcium carbonate and magnesium hydroxide neutralize excess stomach acid, bringing relief from heartburn.

Applications:

Industry: Acids and bases are essential in manufacturing fertilizers, pharmaceuticals, and plastics, like versatile instruments in a chemical orchestra.
Food: Acids preserve and flavor our food, while bases help create baking soda and other additives, adding taste and texture to our meals.
Cleaning: Acids and bases are the cleaning crew, removing dirt and grime from our homes and workplaces.

Medicine: From antacids to chemotherapy, acids and bases play vital roles in treating various ailments.
Agriculture: Alkalis help adjust soil pH, ensuring healthy growth for our crops.

5. The Grand Finale

Acids, bases, and alkalis are fundamental notes in the grand symphony of chemistry. Their interactions shape our world, from the food we eat to the environment we live in. By understanding their properties and behaviors, we can better appreciate the intricate chemical dance that surrounds us and harness their power for the benefit of humankind.

Acids: The Sour Powerhouses of Chemistry

Imagine a world without the tangy zest of a lemon, the invigorating fizz of a soda, or the eye-opening sourness of vinegar. These sensations, we owe to acids, the tiny chemical powerhouses that pack a punch!

Acids are like the superheroes of the molecular world, always ready to react and transform. They're the reason your car battery works, why your stomach digests food, and how your blood maintains a healthy balance.

Metals Beware!

Acids have a particular dislike for metals. When they meet, it's like a tiny wrestling match, with the acid always winning. The acid grabs onto the metal, tearing it apart and releasing hydrogen gas, a lightweight champion that escapes into the air. This is why acids are used to clean metals, stripping away rust and grime like a chemical superhero.

Neutralizing the Villains

But acids aren't all about destruction. They also play a crucial role in maintaining balance. When they encounter a base, their arch-nemesis, they engage in an epic battle known as neutralization. The result? A peaceful truce, forming water and a salt, like a chemical handshake.

This neutralization reaction is used in everything from treating heartburn to cleaning up wastewater. It's like a chemical peace treaty, ensuring harmony in the world of molecules.

The Fizz Factor

Acids also have a knack for creating excitement. When they meet a carbonate, it's like a party in a test tube, with bubbles of carbon dioxide gas fizzing and popping. This is why baking soda is a must-have in the kitchen, helping cakes rise and cookies become fluffy.

Bases: The Slippery Peacekeepers

If acids are the superheroes, then bases are the peacekeepers. They're the cool, calm, and collected members of the chemical world, always ready to neutralize the acidic chaos.

Bases are like the gentle giants of chemistry. They're slippery to the touch and have a bitter taste, but they're essential for life. They're found in everything from soap to antacids, keeping us clean and our stomachs happy.

Neutralizing the Acidic Threat

Bases are the ultimate peacemakers, always ready to neutralize the acidic villains. They're like the chemical equivalent of a warm hug, calming the acidic storm and restoring balance.

This neutralization reaction is crucial in many industrial processes, like treating wastewater and producing fertilizers. It's like a chemical peacekeeping mission, ensuring a harmonious balance in the world of molecules.

The Ammonia Connection

Bases also have a secret weapon: they can release ammonia gas, a pungent compound that's essential for life. This reaction is used to produce fertilizers, which help plants grow and provide us with food.

The Chemical Dance

Acids and bases are like two sides of the same coin, constantly interacting and balancing each other out. They're the yin and yang of chemistry, the dynamic duo that keeps the world of molecules in harmony.

So next time you bite into a tangy lemon or feel the relief of an antacid, remember the amazing chemical dance of acids and bases, the tiny powerhouses that make our world so vibrant and diverse.

Imagine a Chemical Tango: Acids and Bases in a Neutralization Dance

Let's dive into the fascinating world of chemistry, where molecules are always on the move, interacting and reacting in a mesmerizing dance. Today, we'll focus on a particular dance move called "neutralization," where two seemingly opposite partners, acids and bases, come together in a harmonious embrace.

Acids: The Sour-Faced Dancers

Acids, oh those sour-faced dancers! They're known for their tangy taste, like the zing in a lemon or the sourness of vinegar. But their true character lies in their ability to donate a tiny but mighty particle, the hydrogen ion (H+). Think of it as a little spark of energy they carry around.

Bases: The Smooth Operators

On the other side of the dance floor, we have the bases, the smooth operators. They're the opposite of acids, often bitter in taste and slippery to the touch, like soap. Their secret weapon? The hydroxide ion (OH-), a graceful molecule ready to accept the spark of energy from the acids.

The Neutralization Tango: A Balancing Act

When acids and bases meet, they engage in a captivating neutralization tango. The hydrogen ions from the acid and the hydroxide ions from the base come together, like two puzzle pieces fitting perfectly, to form a molecule we all know and love: water (H2O).

The pH Scale: A Dance Floor Guide

To keep track of the action on the dance floor, we use the pH scale, a handy guide that tells us how acidic or basic a solution is. It's like a mood meter, ranging from 0 to 14. Acids have a low pH, making them the grumpy dancers with scores below 7. Bases, with their cheerful disposition, have a high pH, scoring above 7. And when the dance floor is perfectly balanced, with an equal number of acids and bases, we have a neutral solution with a pH of 7.

Indicators: The Dance Floor Judges

To help us follow the neutralization tango, we have special guests called indicators. These molecules change color depending on the pH of the solution, acting like judges on the dance floor. Litmus, for example, turns red in the presence of acids and blue in the presence of bases. It's like a mood ring for the dance floor!

The Importance of the Neutralization Dance

This chemical tango isn't just a spectacle; it plays a vital role in our world. In our bodies, it helps maintain a healthy balance, neutralizing excess stomach acid. In the environment, it helps remediate acidic lakes and soils caused by acid rain. And in industries, it's used in wastewater treatment and food production.

Neutralization: A Chemical Dance of Balance

So next time you encounter the term "neutralization," remember the captivating dance of acids and bases. It's a reminder that even seemingly opposite forces can come together in a harmonious balance, creating something new and essential.

Acids: The Tiny Titans That Pack a Punch (and a Whisper)

Imagine the chemical world as a bustling city. In this metropolis, acids are the busybodies, the movers and shakers, involved in countless reactions and processes. Some acids are like roaring lions, loud and powerful, while others are more like quiet but persistent moles, subtly shaping their surroundings.

Strong Acids: The Chemical Lions

These are the heavy hitters of the acid world, the ones that make a splash wherever they go. Think of hydrochloric acid (HCl), the potent stomach acid that helps you digest that giant burrito. Or sulfuric acid (H2SO4), the workhorse of the chemical industry, found in everything from car batteries to drain cleaner (though hopefully not in the same bottle!).

When these acids hit water, it's like a wild party. They completely break apart, or dissociate, into their ions, releasing a flood of hydrogen ions (H+). These H+ ions are what give acids their characteristic punch, their ability to react with other substances and make things happen.

Weak Acids: The Persistent Moles

Don't let their name fool you, weak acids aren't necessarily wimpy. They're just more subtle in their approach. Think of acetic acid, the gentle giant that gives vinegar its tang. Or carbonic acid, the fizzy friend that makes soda pop and keeps your blood at the right ph.

When weak acids meet water, they're a bit more reserved. They only partially dissociate, releasing a smaller number of H+ ions. They're like the quiet but persistent moles, slowly but surely shaping the chemical landscape.

Why Does Strength Matter?

The strength of an acid, whether it's a roaring lion or a persistent mole, determines its personality and how it interacts with the world. Strong acids are more corrosive and reactive, while weak acids are gentler and more subtle.

This difference in strength is crucial in all sorts of areas:

Chemistry: Knowing how strong an acid is helps chemists predict how it will behave in reactions, like whether it will eat through metal or just give a gentle fizz.
Biology: Our bodies are finely tuned machines, and the strength of acids plays a vital role in keeping everything running smoothly. For example, the weak acid carbonic acid helps maintain the delicate pH balance of our blood.
Industry: From making fertilizers to producing plastics, acids are essential ingredients in countless industrial processes. The strength of the acid determines which jobs it's best suited for.
Environment: Acid rain, caused by strong acids like sulfuric and nitric acid, can wreak havoc on ecosystems and infrastructure. Understanding acid strength helps us tackle these environmental challenges.

The Acid Strength Spectrum

It's not just a black-and-white world of strong and weak. Acids fall on a spectrum of strength, from the super-strong super acids to the mild-mannered weaklings. Scientists are constantly exploring this spectrum, discovering new acids and finding new ways to measure their strength.

The Future of Acids

The world of acids is constantly evolving. Scientists are developing new super acids with incredible strength and exploring the potential of ionic liquids, a new class of acidic materials. With the help of computational chemistry, we're gaining a deeper understanding of what makes acids tick and how we can harness their power for good.

So next time you encounter an acid, whether it's in a lemon, a battery, or a chemistry lab, remember that it's not just a sour-tasting substance. It's a tiny titan with a unique personality, playing a vital role in the grand symphony of chemistry.

Weak Acids: A Quirky Chemist's Guide (with Cartoon Cameos!)

Hey there, fellow science explorers! Today, we're diving deep into the wacky world of weak acids. Imagine them as those laid-back molecules in the chemistry party who don't like to make a fuss. Unlike their rowdy cousins, the strong acids (who throw all their protons around like confetti), weak acids like to keep a few protons close.

What Makes Weak Acids So... Weak?

Think of it like this: strong acids are like those over-eager puppies who can't wait to give you their slobbery tennis ball (that's the proton!). Weak acids, on the other hand, are more like that cool cat who might grace you with a head bop (a proton donation) if they're in the mood.

In other words, weak acids only partially dissociate in water. They're a bit hesitant to let go of their precious protons, creating a dynamic equilibrium – a constant push and pull between the acid molecule and its ions.

Introducing the Weak Acid Crew (with Special Guest Appearances!)

Let's meet some of the most interesting characters in the weak acid club:

The Carboxylic Acid Family:
Acetic Acid (CH_3COOH): The sourpuss of the group, found in vinegar. Think of him as a grumpy old chef who adds a zing to your salads!

(Imagine a cartoon of a vinegar bottle with a grumpy face)

$CH_3COOH(AQ) \rightleftharpoons H^+(AQ) + CH_3COO^-(AQ)$

Formic Acid (HCOOH): This sneaky fellow is found in ant and bee stings. He's small but packs a punch!
(Imagine a cartoon ant with a tiny vial of acid)

$HCOOH(AQ) \rightleftharpoons H^+(AQ) + HCOO^-(AQ)$

Citric Acid ($C_6H_8O_7$): The life of the party, found in citrus fruits. This energetic molecule keeps our cells powered up!
(Imagine a cartoon lemon with a superhero cape)

Lactic Acid ($C_3H_6O_3$): The fitness fanatic. This one builds up in your muscles after a good workout, making them feel like they're on fire!
(Imagine a cartoon muscle flexing and sweating)

Oxalic Acid ($C_2H_2O_4$): The cleaning crew member. Found in plants like spinach, this acid helps keep things sparkling clean (and can even bleach your clothes!).
(Imagine a cartoon spinach leaf with a cleaning spray bottle)

Other Notable Weak Acids:
Carbonic Acid (H_2CO_3): The bubbly one. This acid forms when carbon dioxide dissolves in water, like in your fizzy drinks! It also helps transport CO2 in your blood.
(Imagine a cartoon of a soda bottle with bubbles coming out)

$H_2CO_3(AQ) \rightleftharpoons H^+(AQ) + HCO_3^-(AQ)$
Phosphoric Acid (H_3PO_4): The generous one. This acid can donate three protons! It's found in fertilizers and even some sodas.
(Imagine a cartoon of a phosphate molecule juggling three protons)

Hydrofluoric Acid (HF): The artist. This acid is used to etch glass, creating beautiful designs.
(Imagine a cartoon of a bottle of acid etching a design on a window)

Hydrogen Sulfide (H_2S): The stinky one. This gas has the distinct aroma of rotten eggs. Hold your nose!
(Imagine a cartoon of a rotten egg with a smelly cloud around it)

Universal Indicator: The pH Detective

Now, how do we measure how acidic or alkaline something is? Enter the universal indicator! This colorful character changes color depending on the pH of a solution. It's like a mood ring for chemicals!

(Imagine a cartoon of a universal indicator strip with a magnifying glass, like a detective)

Aquarium Adventures: A pH Balancing Act

Ever wondered how to keep your pet fish happy? Maintaining the right pH in their aquarium is crucial! Universal indicator can help you become a pH detective for your fishy friends, ensuring their water is just right.

(Imagine a cartoon fish swimming happily in a tank with a universal indicator strip showing the perfect pH)

Wrapping Up the Acidic Adventure

So, there you have it – a whirlwind tour of the world of weak acids! These fascinating molecules play a vital role in everything from the food we eat to the environment around us. And with the help of the universal indicator, we can keep track of their activity and ensure a healthy balance in our world.

The Chameleon Chemistry of Amphoteric Oxides

Imagine a substance that can't quite make up its mind. Like a chameleon adapting to its surroundings, it can shift its behavior depending on the company it keeps. That's the intriguing world of amphoteric oxides – the rebels of the chemical world that refuse to be confined to a single identity.

These fascinating compounds possess a "dual personality," capable of acting as both an acid and a base. It's like having a superhero with two alter egos, each ready to tackle a different villain (or in this case, react with a different type of chemical).

Why the Split Personality?

This Jekyll and Hyde behavior stems from their unique position in the chemical landscape. They're formed from elements that straddle the line between metals and non-metals, giving them a balanced nature that allows them to react with both acidic and basic substances.

A Tale of Two Reactions

Picture aluminum oxide (Al2O3), a classic amphoteric oxide, as our protagonist. When it encounters a strong acid like hydrochloric acid (HCl), it transforms into a base, accepting protons and forming a salt and water. But when it faces a strong base like sodium hydroxide (NaOH), it switches roles, acting as an acid and donating protons to form a different salt and water.

Beyond the Textbook

Amphoteric oxides aren't just theoretical curiosities confined to textbooks. They play vital roles in various industries:

Aluminum oxide, with its high melting point and hardness, is a workhorse in ceramics, refractories, and catalysis.
Zinc oxide is a common ingredient in sunscreen lotions and ointments, thanks to its ability to absorb UV radiation and fight bacteria.

A Chemical Balancing Act

The amphoteric nature of these oxides highlights the delicate balance between acidity and basicity in the chemical world. It's a reminder that even in the seemingly rigid realm of chemistry, there's room for flexibility and adaptability. So next time you encounter an amphoteric oxide, remember its chameleon-like character – a testament to the fascinating complexity of the molecular world.

Preparation of salts

The Salt Shaker's Guide to Crystal Creation: A Chemist's Cookbook

Forget baking cakes, we're cooking up crystals! Dive into the fascinating world of soluble salts, those magical compounds that dissolve in water like sugar in tea. But how do we conjure these crystalline creations? Fear not, fellow alchemists, for here's a whimsical guide to preparing, separating, and purifying these shimmering wonders.

Part 1: The Alchemy of Acids and Bases

Imagine acids and bases as two rival kingdoms, constantly battling for control. When they meet in the right proportions, they neutralize each other, giving birth to a peaceful salt kingdom. This is the essence of titration, a delicate dance where we carefully add acid to a base (or vice versa) until perfect harmony is achieved.

Think of it like adding lemon juice to your tea: too much, and it's sour; too little, and it's bland. But just the right amount, ah, that's the sweet spot! We use an indicator, a magical substance that changes color to signal the end of the battle, ensuring our salt solution is perfectly balanced.

Part 2: Unveiling the Crystalline Treasures

Now that our salt solution is ready, it's time to unveil the hidden crystals. Gently heating the solution evaporates the water, leaving behind a concentrated treasure trove. As the water disappears, the salt ions huddle closer, eventually locking arms to form a dazzling array of crystals.

Picture a bustling dance floor where everyone's moving freely. As the music slows and the lights dim, couples pair up and sway together, forming a beautiful pattern on the floor. That's our crystallization in action!

Part 3: Polishing the Gems

Our crystals might still be clinging to some unwanted impurities, like party crashers at our grand ball. To evict these unwanted guests, we use filtration, a sieve that separates the solid crystals from any remaining liquid.

Finally, we give our crystals a refreshing bath with distilled water and gently dry them, like polishing precious gems to reveal their full brilliance.

A World of Crystalline Wonders

From the soothing Epsom salts to the light-sensitive silver chloride used in photography, soluble salts play a vital role in our lives. By mastering these techniques, we unlock the secrets of creating these versatile compounds.

So, don your lab coats and embrace your inner alchemist. The world of salt crystals awaits your creative touch!

Bonus: Insoluble Salts - The Rebels

Not all salts are eager to mingle with water. Some prefer to stay solid, forming precipitates that cloud the solution. Think of them as the rebels who refuse to conform. To create these insoluble salts, we mix two soluble solutions, each containing one half of the desired rebel pair. When these solutions meet, the rebel ions find each other and, in a dramatic exit, form a solid precipitate.

Beyond the Basics

This is just the beginning of our crystal-creating journey. There's a whole universe of salts out there, each with unique properties and applications. So, keep experimenting, keep exploring, and who knows? Maybe you'll discover a brand-new crystal with magical properties!

Imagine a bustling city: skyscrapers reaching for the clouds, intricate networks of roads connecting every corner, and a constant flow of energy keeping everything alive. Now, picture this city on a molecular level. That's the world of hydrated substances – compounds where water molecules aren't just visitors, they're vital citizens, woven into the very fabric of the city's structure.

These water molecules, the "water of crystallization," aren't just hanging around; they have designated addresses within the crystal lattice, like residents with reserved parking spots. They play a crucial role in maintaining the city's stability, influencing its shape, color, and even its social life (reactivity).

Think of copper (II) sulfate pentahydrate as a glamorous celebrity with an entourage of five water molecules (Cuso4•5H2O). Without them, it's just plain old "anhydrous" copper (II) sulfate, a wallflower in a white dress. But add those water molecules, and it transforms into a dazzling blue crystal, ready to hit the red carpet.

Why does this matter?

Well, imagine trying to build a house without cement. Water of crystallization is like the cement in these crystal cities, holding everything together. It affects how the city functions, how it interacts with its environment, and even how it looks.

Real-world examples:

Gypsum (Caso4•2H2O): This workhorse is a key ingredient in construction materials, like the drywall in your home.
Epsom Salt (Mg SO4•7H2O): This multi-talented compound can soothe sore muscles, fertilize your garden, and even help you relax in a bath.

Washing Soda ($Na_2CO_3 \cdot 10H_2O$): A cleaning powerhouse that helps keep your clothes sparkling and your dishes spotless.
Hydrated compounds are like chameleons, changing their properties depending on their water content. Cobalt (II) chloride hexahydrate, for instance, shifts from a playful pink when hydrated to a serious blue when anhydrous. This color-changing ability makes it a star ingredient in humidity indicators and desiccants.

The takeaway?

Water of crystallization is more than just a fancy term. It's a fundamental concept that impacts everything from chemistry and medicine to environmental science and food technology. By understanding how these "water cities" function, we can unlock new possibilities in materials science, drug development, and even water purification.

So next time you encounter a crystal, take a moment to appreciate the hidden world within, a bustling metropolis of molecules where water plays a starring role.

Imagine the Salt Shaker of Life

Imagine a salt shaker. Not just any salt shaker, but the Salt Shaker of Life. Inside, aren't just boring old white crystals. It's filled with a rainbow of salts, each with its own personality and quirks. Some are eager to dive into water, dissolving instantly like excited puppies jumping into a lake. Others are more hesitant, clinging to each other like shy wallflowers at a party. And then there are the rebels, stubbornly refusing to mix with water no matter what.

These "personalities" are what we call solubility rules. They're like social guidelines for salts, helping us predict who's going to mingle with water and who's going to stay aloof.

The Life of the Party

Sodium, Potassium, and Ammonium: These guys are the life of the party, always the first ones on the dance floor (or in the water, in this case). They're so eager to join the water molecules that they practically burst with excitement, leaving their crystal structure behind without a second thought. Think of them as the popular kids in school, everyone wants to be their friend.

Nitrates: These are the cool kids, the ones who are effortlessly popular without even trying. They're large and in charge, with a "been there, done that" attitude. Water molecules can't resist their charm, and neither can we.

The Wallflowers

Chlorides: Most chlorides are happy to mingle, but they have a few shy friends who prefer to stick to themselves. Lead and silver chloride are like those kids who always hang back in the corner, whispering secrets and avoiding eye contact.

Sulfates: Sulfates are generally outgoing, but they have a few close friends they can't bear to be separated from. Barium, calcium, and lead sulfate are like the inseparable trio who always show up together, finish each other's sentences, and never leave each other's side.

The Rebels

Carbonates: These are the rebels, the non-conformists who refuse to follow the crowd. Most carbonates are like grumpy old men, shaking their fists at the water and muttering about "kids these days." But there are a few exceptions, of course. Sodium, potassium, and ammonium carbonate are the young rebels, the ones who break the rules just to prove they can.

Hydroxides: Hydroxides are like the mysterious loners, always lurking in the shadows and avoiding social interaction. Most of them are completely insoluble, like hermits living in caves. But a few, like sodium, potassium, ammonium, and partially calcium hydroxide, are more like the misunderstood artists, hiding their true beauty from the world.

Beyond the Salt Shaker

These solubility rules aren't just fun and games. They're essential for understanding how our world works. They explain why the ocean is salty, why some rocks dissolve in rainwater, and why certain medicines work the way they do. So next time you sprinkle salt on your food, take a moment to appreciate the complex social lives of these tiny crystals. You might be surprised at what you discover.

The Periodic Table

Arrangement of elements

The Periodic Table: A Quirky Family Reunion

Imagine the periodic table as a massive family reunion, with each element a unique individual with its own personality quirks and family relationships. Some are loud and reactive, like the alkali metals, always eager to make a splash. Others are quiet and reserved, like the noble gases, content in their own little world.

1. Finding Your Place at the Reunion

This family gathering is organized in a very specific way:

Periods (Rows): Think of these as the different generations of the family. Each row represents a new generation with a larger electron shell, like adding a new floor to the family mansion.
Groups (Columns): These are the family cliques, the groups of cousins who share similar interests (chemical properties) because they have the same number of valence electrons – the "social butterflies" of the atom.
Atomic Number: This is like the family roll call, starting with hydrogen (number 1) and going all the way down the line. Each number represents the number of protons in an element's nucleus, its unique family ID.

2. From Wallflowers to Life of the Party

As you move across a period, the elements gradually change from shy wallflowers (metals) to the life of the party (non-metals). This is because their personalities (properties) change:

Electronegativity: How strongly an element clings to electrons, like someone guarding their favorite dessert. Non-metals are the ultimate hoarders.

Ionization Energy: The energy needed to pry an electron away, like convincing a grumpy teenager to leave their room. Non-metals are notoriously stubborn.
Electron Affinity: How much an element wants to gain an electron, like someone eager to join a popular club. Non-metals are always recruiting.
Atomic Radius: The size of the atom, like the personal space an element need. Non-metals are a bit more compact.

3. Family Cliques and Their Quirks

The group number tells you a lot about an element's social behavior:

Groups 1, 2, and 13: These are the generous types, always ready to lose an electron (like lending a helping hand) to become positive ions.
Groups 15, 16, and 17: The collectors, eager to gain electrons (like acquiring rare stamps) to become negative ions.
Group 14: The ambiverts, comfortable both giving and taking electrons, depending on the situation.
Group 18: The introverts, the noble gases, perfectly content with their full outer shell of electrons, like a complete puzzle. They rarely interact with others.

Case Study: The Alkali Metal Mayhem

Imagine the alkali metals (Group 1) as a group of hyperactive toddlers, each with one toy (valence electron) they're eager to share (lose). This makes them super reactive, especially around water:

Lithium (Li): Dips its toe in the water and creates a little ripple.
Sodium (Na): Jumps in with a splash, making a bigger commotion.
Potassium (K): Does a cannonball, causing a huge wave!
The further down the group you go, the wilder the reaction, like the toddlers getting more energetic as the day goes on.

In Conclusion

The periodic table is like a family portrait, capturing the unique personalities and relationships of all the known elements. By understanding its organization and trends, we can unlock the secrets of this fascinating family and appreciate the incredible diversity of our universe.

The Periodic Table: A Quirky Family Reunion of Elements

Imagine the periodic table as a massive family reunion, where each element is a unique individual with its own quirks and personality. Instead of boring old rows and columns, picture a sprawling picnic blanket where elements gather in clusters based on their shared traits and family resemblances.

1. "Like Elements Stick Together" - The Power of the Clique

Just like in any big family, elements with similar characteristics tend to form cliques. These cliques, or groups as chemists call them, are where the real action happens. The secret to their bond? Their "valence electrons" – the social butterflies of the atomic world, always eager to mingle and form connections.

Think of the alkali metals (Group 1) as the hyperactive kids in the family, always bouncing off the walls with their single valence electron. They're desperate to share this electron, making them super reactive and prone to forming strong bonds (sometimes explosively!).

Then there are the halogens (Group 17), the moody teenagers of the periodic table. With seven valence electrons, they're just one shy of a full "shell" and will do almost anything to get that last electron, making them equally reactive, but in a “grabbier” way.

And let's not forget the noble gases (Group 18), the aloof grandparents sitting in the shade. They've already got a full shell of electrons, so they're content and unreactive, happy to observe the chaos from a distance.

2. "Location, Location, Location!" - Predicting an Element's Personality

The beauty of this family reunion is that you can predict an element's personality based purely on where it's sitting on the picnic blanket.

Think of the periods (horizontal rows) as generational lines. As you move across a period, from left to right, the elements get smaller (like kids shrinking as they age!), become more electronegative (clingier!), and less metallic (less shiny and exciting).

Moving down a group (vertical column) is like tracing a family lineage. Elements get bigger (like growing taller!), less electronegative (more independent!), and more metallic (more dazzling!).

For example, chlorine (Cl) is a feisty halogen teenager sitting towards the top right corner of the blanket. You can bet it's going to be highly electronegative (a real attention seeker!), reactive, and smaller than its older sibling, bromine (Br).

3. "Family Traits" - Spotting the Trends

Just like families share physical traits, elements in the same group exhibit similar trends in their behavior.

The "Wild Child" Trend: Alkali metals get wilder (more reactive) as you move down the group. Think of it as the younger siblings rebelling even harder!
The "Mellowing Out" Trend: Halogens become less electronegative (less clingy) as you go down the group. They've learned to be more independent!
The "Stronger Bonds" Trend: Noble gases get harder to boil (stronger bonds) as you move down the group. It's like their family ties are getting tighter with age.

Beyond the Picnic Blanket: The Ever-Growing Family

This periodic table picnic is constantly expanding with new elements joining the party. These newcomers, like nihonium, meconium, and tennessine, are the mysterious distant relatives with surprising properties that challenge our understanding of the family dynamics.

So, the next time you glance at the periodic table, don't just see a boring chart. Imagine it as a lively family reunion, full of fascinating characters and intriguing relationships. It's a roadmap to understanding the elements, their personalities, and how they interact to create the world around us.

Group I properties

The Alkali Metal Crew: A Family Drama

Imagine the periodic table as a bustling apartment complex. In one corner, you have the Alkali Metal family, a group of siblings known for their dramatic personalities and explosive tendencies.

Lithium, the Eldest:

Personality: A bit of a loner, Lithium keeps to himself. He's got a reputation for being a bit moody, sometimes reacting calmly, other times throwing a tantrum when things get too hot.
Claim to Fame: He's the lightweight of the family, often found chilling in your phone battery, keeping the energy flowing.

Sodium, the Middle Child:

Personality: Sodium craves attention. He's loud, flamboyant, and loves to make a splash, especially when water is involved.
Claim to Fame: He's the life of the party, lighting up the streets with his bright yellow glow in those old-school streetlamps.

Potassium, the Rebel:

Personality: Potassium is the wild child, always pushing the boundaries. He's got a fiery temper and a tendency to overreact, especially when someone tries to control him.
Claim to Fame: He's a bit of a health nut, hanging out in fertilizers and helping plants grow big and strong.

Family Dynamics:

These siblings have a lot in common: they're all silvery-smooth and love to shed their single "valence electron" jacket like it's going out of style. But they also have their differences. As you move down the family line, things get more intense:

Melting Point Drama: Lithium is a homebody, preferring to stay solid until things get really heated. But his younger brothers have a serious case of wanderlust, melting at lower temperatures.
Density Debacle: Sodium and Potassium are constantly arguing about who's the densest. Sodium usually wins, but Potassium has a secret weapon: his larger size sometimes makes him less dense, throwing a wrench in the competition.
Reactivity Rumble: When it comes to reactions, things get explosive. Lithium might simmer, but Sodium and Potassium go off with a bang, especially when water is around. They're always competing to see who can react the fastest with oxygen and halogens, creating a chaotic scene in the periodic table apartment complex.

The Extended Family:

The Alkali Metal family doesn't end there. They have some mysterious cousins – Rubidium, Cesium, and Francium – who live further down the hall. These relatives are even more reactive and unpredictable, making them the subject of hushed whispers in the periodic table community.

The End... or is it?

The Alkali Metals are a family full of drama, excitement, and a touch of danger. Their unique personalities and trends make them essential players in the world around us, from powering our devices to lighting our streets. So next time you encounter an alkali metal, remember the family drama unfolding within the periodic table.

Group VII properties

The Halogens: A Colorful Cast of Characters

Imagine a family reunion, but instead of quirky aunts and uncles, we have a gathering of elements with vibrant personalities. These are the halogens, a group of siblings from the periodic table known for their dramatic flair and eagerness to react. Today, we're shining the spotlight on three of them: chlorine, bromine, and iodine.

1. Family Traits and Quirks

Halogens are like those close-knit families where everyone shares a striking resemblance. They all exist as pairs (diatomic molecules), and they're always on the lookout for an extra electron to complete their outer shell. This makes them highly reactive, like that friend who always jumps into a conversation without a second thought.

But even within families, there are differences:

Density Drama: As we move down the family line, from chlorine to iodine, things get denser. Imagine them as a stack of pancakes, with iodine being the thickest and most substantial one at the bottom.
Reactivity Rollercoaster: While all halogens are eager to react, their enthusiasm wanes as we go down the group. Chlorine is like the hyperactive youngest sibling, while iodine is the more laid-back older one. This is because the larger the atom, the weaker its grip on incoming electrons.

2. A Visual Feast

Halogens are not just chemically interesting; they're also a treat for the eyes:

Chlorine: A pale yellow-green gas that would fit right into a sci-fi movie. It has a pungent smell that screams, "Don't get too close!"
Bromine: A dense, red-brown liquid that seems to smolder with intensity. It easily transforms into a gas, like a magician vanishing in a puff of smoke.

Iodine: A dark, almost mystical solid with a metallic sheen. When heated, it undergoes a dramatic transformation, sublimating into a vibrant purple gas. It's like witnessing a shy creature suddenly burst into a colorful dance.

3. Meet the Stars

Let's get to know each of these elements a bit better:

Chlorine: This element has a history of making a splash. Discovered in 1774, it was initially mistaken for something else entirely. Today, it's a busybody involved in everything from keeping our water clean to making plastics.
Bromine: Discovered in 1826, this element is less abundant but no less important. It helps keep things safe by acting as a flame retardant and even plays a role in photography and medicine.
Iodine: Discovered in 1811, this element is a bit of a health nut. It's essential for our thyroid hormones and is often added to salt to prevent deficiencies. It also has a knack for medical imaging and acts as a catalyst in chemical reactions.

4. Real-World Tales

Halogens aren't just confined to the lab; they're out in the world, making a difference:

Chlorine, the Water Guardian: Thanks to chlorine, we can enjoy clean and safe drinking water, preventing the spread of waterborne diseases.
Bromine, the Safety Officer: Brominated flame retardants help prevent fires, but they also come with environmental concerns, reminding us that even helpful elements can have a dark side.
Iodine, the Essential Nutrient: Iodine deficiency can lead to health problems, highlighting the importance of this often-overlooked element.

5. Curtain Call

Chlorine, bromine, and iodine, each with its unique personality and role to play, demonstrate the diversity within the halogen family. Their story is a reminder that even the smallest elements can have a big impact on our lives and the world around us.

Imagine a chemical dance floor... where the halogen elements are the stars, each vying for a coveted spot under the disco ball. These elements, with their insatiable desire for electrons, are always ready to cut in and steal the partner of another. This, my friend, is the essence of a displacement reaction!

Picture fluorine, the feisty dancer, small yet incredibly energetic. It's the life of the party, always ready to snatch an electron from any other halogen. Chlorine, a bit larger and less agile, can still hold its own, out-dancing bromine and iodine. Bromine, with its reddish-brown glow, is like the smoldering, seductive dancer, while iodine, the gentle giant, prefers to observe from the sidelines.

Now, imagine chlorine swirling across the dance floor towards bromine. A dramatic exchange takes place! Chlorine, with its stronger attraction to electrons, steals bromine's partner, leaving bromine to sulk alone. The dance floor changes color as bromine's reddish hue intensifies, a testament to the chemical tango that just occurred.

But the drama doesn't stop there! Bromine, seeking revenge, glides towards iodine, who's shyly holding an electron. With a swift move, bromine snatches the electron, leaving iodine electron-less and heartbroken. The dance floor shifts again, this time turning a deep brown as iodine reveals its true colors.

This captivating dance of electrons is not just a spectacle; it's a fundamental principle of chemistry. It's the reason why chlorine can purify water by displacing less reactive elements from bacteria, and it's the key to understanding the behavior of these fascinating elements.

Think of the periodic table as a VIP guest list, with each element assigned a unique ranking based on its reactivity. Fluorine, at the top of the list, always gets the first dance, while tennessine, the mysterious newcomer at the bottom, might be too shy to even step onto the floor.

So, the next time you encounter a halogen displacement reaction, don't just see it as a chemical equation. Imagine the vibrant dance of electrons, the dramatic exchange of partners, and the ever-changing colors on the chemical dance floor. It's a story of attraction, competition, and the fundamental laws of nature, all playing out in a beaker!

Transition elements

Transition Elements: The Stars of the Periodic Table

Imagine the periodic table as a grand stage, and the elements as its cast of characters. Enter the transition elements, the captivating stars of the show! These aren't your everyday, run-of-the-mill elements; they're the divas, the scene-stealers, with a flair for the dramatic and a knack for the unexpected.

Why all the fuss? Well, it's all thanks to their unique electronic configurations, specifically those partially filled d orbitals. Think of these orbitals as backstage dressing rooms where electrons can change their outfits (energy levels) and emerge with dazzling new colors and abilities.

Let's pull back the curtain and shine a spotlight on some of their most show-stopping properties:

(a) Density: Heavyweight Champions of the Elements

These metals are the bodybuilders of the periodic table, packing more mass into the same space than their alkali and alkaline earth metal counterparts. Picture them as tiny, super-dense marbles, thanks to:

Strong Metallic Bonding: Imagine a bustling city with a vast network of interconnected roads (bonds) and skyscrapers (atoms). Transition metals have a massive network of these bonds, making them incredibly strong and tightly packed.
High Atomic Mass: Like a star with a massive gravitational pull, transition metals have a high atomic mass, drawing their atoms closer together.

Lanthanide Contraction: This phenomenon is like a cosmic shrinking ray, pulling in the atoms of the second and third transition series, making them even denser.

Star Performers:

Osmium (OS): The undisputed heavyweight champion, so dense it would feel like holding a miniature black hole!
Iridium (IR): A close runner-up, nearly as dense as osmium and equally impressive.
Platinum (Pt): The glamorous star, adored for its beauty and resilience in jewelry and as a catalyst.

Case Study: Tungsten: The Tough Guy

Tungsten, with its incredible density, is the go-to element for extreme jobs:

Piercing Power: Imagine a bullet made of tungsten, so dense it can pierce even the toughest armor.
Radiation Shield: Like a superhero deflecting harmful rays, tungsten protects us from radiation in medical and industrial settings.
Precision Balance: In Formula 1 cars and aircraft, tiny tungsten weights ensure perfect balance, like a tightrope walker's pole.

(b) Melting Points: Unflinching in the Face of Heat

These metals laugh in the face of high temperatures, refusing to melt easily. Their secret?

Strong Metallic Bonding: Those tightly packed atoms and strong bonds create an unbreakable network, requiring immense energy to break apart.
Efficient Packing: Think of perfectly stacked bricks forming a sturdy wall. The atoms in transition metals are arranged in a similarly efficient way, maximizing their strength.

Star Performers:

Tungsten (W): The ultimate heat-defer, with a melting point that would make even the hottest stars sweat.
Rhenium (Re): A close second, almost as resistant to melting as tungsten.
Tantalum (Ta): Used in high-temperature applications, like a spaceship's heat shield, withstanding the fiery re-entry into Earth's atmosphere.

Case Study: Iron: The Backbone of Industry

Iron, with its impressive melting point, is the workhorse of the steel industry:

Blast Furnace Inferno: Iron ore is melted down in a fiery blast furnace, like a blacksmith's forge, to produce molten iron.
Steelmaking Magic: The molten iron is transformed into steel, adding ingredients and controlling the carbon content, all while withstanding scorching temperatures.
Shaping the World: Molten steel is poured into molds to create everything from skyscrapers to cars, thanks to its ability to flow and solidify at high temperatures.

(c) Variable Oxidation States: Masters of Disguise

Transition metals are the chameleons of the elemental world, changing their appearance (oxidation states) with ease. This is due to:

Multiple Electron Loss: Like a quick-change artist shedding layer of clothing, transition metals can lose electrons from different orbitals, leading to a variety of oxidation states.
Ligand Effects: The surrounding molecules (ligands) can influence the metal's oxidation state, like a costume designer dressing the actor for different roles.

Star Performers:

Manganese (Mn): The master of disguise, with a range of oxidation states from -3 to +7, like an actor playing both hero and villain.

Chromium (Cr): Another versatile performer, commonly found in +2, +3, and +6 states, like a character with a complex backstory.
Iron (Fe): Known for its +2 (ferrous) and +3 (ferric) states, like a dynamic duo with distinct personalities.

Case Study: Redox Reactions in Biological Systems

In the intricate dance of life, transition metals play vital roles by switching between oxidation states:

Hemoglobin's Oxygen Delivery: Iron in hemoglobin cycles between Fe^{2+} and Fe^{3+}, like a delivery truck picking up and dropping off oxygen throughout the body.
Cellular Respiration's Powerhouse: Iron and copper in cytochrome c oxidase help generate energy for cells, like a power plant keeping the lights on.
Photosynthesis's Green Engine: Manganese in photosystem II helps split water molecules, like a solar panel harnessing the sun's energy.

(d) Colored Compounds: A Kaleidoscope of Chemistry

Transition metal compounds are the artists of the periodic table, painting the world with vibrant colors. This artistry is due to:

d-d Transitions: Electrons in the d orbitals absorb light and jump between energy levels, like a painter mixing different colors on their palette.
Ligand Field Strength: The surrounding ligands influence the color, like a lighting designer changing the mood of a scene.

Star Performers:

Copper (II) sulfate: The classic blue crystal, like a sapphire gemstone.
Potassium permanganate: A deep purple solution, like a royal robe.
Nickel (II) chloride: Yellow when dry, green when hydrated, like a chameleon changing its colors.

Case Study: Pigments and Dyes: Coloring Our World

For centuries, we've harnessed the beauty of transition metal compounds to create pigments and dyes:

Prussian Blue: A deep blue pigment used in paints and blueprints, like the night sky captured on canvas.
Cadmium Yellow: A bright yellow pigment, once a favorite of artists, now used cautiously due to its toxicity.
Chromium Oxide Green: A stable green pigment found in paints and ceramics, like the lush green of a forest.

(e) Catalytic Activity: The Matchmakers of Chemistry

Transition metals are the ultimate matchmakers, bringing together reactants and speeding up chemical reactions. Their skills come from:

Variable Oxidation States: Like skilled negotiators, they can easily switch between states, facilitating electron transfers and lowering the energy barrier for reactions.
Surface Adsorption: They attract reactants to their surface, like a dance floor bringing molecules together.
Formation of Complexes: They form temporary bonds with reactants, weakening existing bonds and making them more reactive, like introducing shy partners at a party.

Star Performers:

Iron: The catalyst in the Haber process, producing ammonia for fertilizers, like a farmer ensuring a bountiful harvest.
Platinum: Used in catalytic converters, cleaning up car exhaust, like a pollution-fighting superhero.
Nickel: Aids in hydrogenation, transforming unsaturated fats into saturated ones, like a chef perfecting a recipe.

Case Study: Catalysis in the Petroleum Industry

The petroleum industry relies heavily on transition metal catalysts to refine crude oil:

Catalytic Cracking: Breaking down large hydrocarbon molecules, like a sculptor chiseling a masterpiece from a block of marble.
Reforming: Rearranging molecules to improve fuel quality, like a mechanic fine-tuning an engine.
Hydrodesulfurization: Removing sulfur from fuels, reducing pollution, like a cleaner purifying the air.

The Final Act

Transition elements are truly the stars of the periodic table, captivating us with their unique properties and endless applications. As we continue to explore their potential, we can expect even more exciting discoveries and innovations in the future. The show has just begun!

Noble gases

The Lone Wolf Elements: A Tale of Noble Gases

Imagine a group of individuals so content in their own skin, so self-sufficient, that they have zero interest in mingling with others. They're the strong, silent types at the party, perfectly happy in their own company. These are the noble gases, the introverts of the periodic table.

Why So Aloof? The Secret to Their Independence

Their secret lies in their perfectly balanced inner world – a complete set of electrons in their outermost shell. It's like having the perfect puzzle, where every piece fits snugly. This "electronic Zen" makes them incredibly stable and unreactive. While other elements are busy swapping electrons to form bonds (like social butterflies flitting between conversations), the noble gases are content in their solitude.

Monatomic Musketeers: All for One and One is All

This self-sufficiency extends to their physical state as well. They exist as single atoms, like lone wolves roaming the vast expanse of the periodic table. No need to pair up or form cliques; they're perfectly fine on their own.

Invisible and Elusive: Ghosts of the Periodic Table

You won't find these gases easily. They're colorless, odorless, and practically invisible under normal conditions. Like ninjas of the element world, they blend seamlessly into their surroundings.

But Wait, There's More! Breaking the Stereotype

While known for their aloofness, the heavier noble gases, like xenon and krypton, have a rebellious streak. Under extreme conditions, they can be coaxed into forming compounds, proving that even the most introverted can be persuaded to step out of their comfort zone.

The Noble Gases in Action: From Party Tricks to Life-Saving Technologies

Despite their introverted nature, noble gases have found their way into a surprising number of applications:

Party Animals: Neon lights, with their vibrant glow, are a testament to the hidden flair of these gases.
Lifesavers: Helium, in its liquid form, cools the powerful magnets in MRI machines, enabling life-saving medical diagnoses.
Welders' Best Friend: Argon provides an inert atmosphere for welding, ensuring strong, clean joints.
Eye Surgeons' Secret Weapon: Krypton lasers are used in delicate eye surgeries, correcting vision with pinpoint accuracy.

The Noble Gases: A Paradox of Introversion and Impact

The noble gases, with their unique blend of introversion and hidden talents, are a testament to the fact that even the most unassuming individuals can have a profound impact on the world around them. They are the quiet achievers, the unsung heroes, proving that sometimes, it's the lone wolves that shine the brightest.

Metals

Properties of metals

Metals vs. Non-metals: A Tale of Two Worlds
Embark on a thrilling expedition through the contrasting realms of metals and non-metals, where we'll uncover their unique personalities and hidden talents.

Metals: The Shining Stars
Picture a bustling city at night, ablaze with lights and energy. That's the world of metals – vibrant, dynamic, and always ready to conduct!

Thermal and Electrical Conductivity: Metals are the ultimate sharers, readily passing along heat and electricity thanks to their free-spirited electrons. Think of copper wires channeling electricity into your home or aluminum pots efficiently distributing heat on the stove.
Malleability and Ductility: Metals are the chameleons of the material world, effortlessly transforming into sheets or wires. They're the reason we have everything from delicate jewelry to sturdy bridges.
Melting and Boiling Points: Most metals are tough cookies, requiring scorching temperatures to melt or boil. Tungsten, for instance, is used in light bulb filaments because it can withstand extreme heat.
Reactions: Metals are eager to react, especially with acids and oxygen. This explains why iron rusts and why some metals generate hydrogen gas when mixed with acids.

Non-metals: The Quiet Heroes
Imagine a peaceful countryside, where things move at a slower pace. That's the domain of non-metals – calm, stable, and often providing essential support.

Thermal and Electrical Conductivity: Non-metals are the insulators, keeping things cool and preventing electrical surges. Think of the wooden handle of your favorite pan or the rubber gloves protecting electricians.
Malleability and Ductility: Non-metals are typically brittle, but some, like certain plastics, can be molded into various shapes.

Melting and Boiling Points: Non-metals have a wide range of melting and boiling points. Diamond, a form of carbon, has an incredibly high melting point, while water, a simple molecule, melts and boils at much lower temperatures.
Reactions: Non-metals are generally less reactive than metals, but they still play crucial roles in various chemical processes. Oxygen, for example, is essential for combustion.

Metals and Non-metals: A Dynamic Duo
Like two sides of the same coin, metals and non-metals complement each other, contributing to the rich tapestry of our material world.

Space Shuttle: The space shuttle's heat shield relied on non-metal ceramic tiles to protect it from the scorching heat of re-entry, while its sturdy frame was made of heat-resistant metals.
Electrical Wiring: Copper wires carry electricity into our homes, while the plastic coating acts as an insulator, preventing electrical shocks.
Jewelry: Gold's malleability and ductility make it ideal for intricate jewelry, while diamonds, with their exceptional hardness, add sparkle and durability.

In Conclusion
Metals and non-metals, with their contrasting properties, are the unsung heroes of our everyday lives. From the smartphones we use to the buildings we inhabit; they play essential roles in shaping our world. So next time you encounter a metal or non-metal, take a moment to appreciate its unique characteristics and the vital role it plays in our material world.

Uses of metals

Metals: The Unsung Heroes of Our Everyday Lives

Imagine a world without metals. No soaring skyscrapers, no speedy cars, no smartphones buzzing with the latest news. It's hard to picture, isn't it? Metals are the silent workhorses of our modern world, their incredible properties shaping almost every aspect of our lives. From the aluminum that allows airplanes to defy gravity to the copper that brings electricity to our homes, these materials are truly the unsung heroes of our everyday existence.

Aluminum: The Lightweight Champion

Aluminum is like the featherweight boxer of the metal world – light on its feet, yet surprisingly strong. This unique combination makes it perfect for situations where weight is a major concern.

Taking Flight: Think of a majestic airplane soaring through the sky. Now imagine it weighed down by heavy steel. Aluminum's low density allows planes to be lighter, which means they use less fuel and can carry more passengers and cargo. It's a win-win for both the environment and the airline's bottom line!
Powering Our Homes: Those long, thin cables that crisscross the countryside, bringing electricity to our homes? Often, they're made of aluminum. Its lightness reduces the strain on the supporting towers and makes it easier to transport and install these vital power lines.
Keeping Our Food Fresh: Ever wondered how your favorite soda stays fizzy or your leftovers stay safe in the fridge? Thank aluminum! Its natural resistance to corrosion means it won't rust and contaminate your food, keeping it fresh and delicious.

Copper: The Electrical Maestro

Copper is like the conductor of an orchestra, expertly guiding the flow of electricity through our homes and gadgets. Its exceptional conductivity and flexibility make it the undisputed king of electrical wiring.

Wiring Our World: From the tiny circuits in your smartphone to the massive cables that power entire cities, copper ensures that electricity flows smoothly and efficiently. Its ability to be drawn into thin wires without breaking makes it ideal for intricate electronic devices.
Connecting Us: Copper has been at the heart of communication for centuries. From the early telegraph lines that spanned continents to the internet cables that connect us today, copper has played a vital role in bringing people closer together.

Beyond the Everyday

But the story of metals doesn't stop there. They're also essential in countless other ways, often behind the scenes:

Titanium: This super-strong, lightweight metal is used in everything from medical implants to jet engines.
Gold: Its beauty and rarity make it a prized possession, but gold also has important industrial uses, thanks to its excellent conductivity and resistance to corrosion.
Steel: The backbone of modern construction, steel's strength and versatility allow us to build towering skyscrapers and massive bridges.

The Future of Metals

As technology advances, we're constantly discovering new ways to use metals and improve their properties. Scientists are developing innovative alloys that are even stronger, lighter, and more resistant to corrosion. These advancements will undoubtedly lead to even more exciting applications for metals in the future, shaping the world we live in for generations to come.

So, the next time you see a gleaming skyscraper, a sleek airplane, or a simple copper wire, take a moment to appreciate the incredible properties of metals and the vital role they play in our lives. They truly are the unsung heroes of our modern world.

Alloys and their properties

Alloys: Where Metals Get Together and Make Something Amazing

Imagine the periodic table as a grand party, with each element showing off its unique personality. Some are shy and retiring, like pure gold, content to shine on their own. Others are boisterous and eager to mingle, like iron or copper. When these social butterflies meet, they form exciting partnerships called alloys.

Alloys aren't just random mixtures; they're carefully orchestrated collaborations where metals (and sometimes a non-metal or two) come together to create something entirely new and often far superior to their individual selves. It's like a superhero team-up – each member brings their strengths to the table, resulting in a powerful combination that can tackle challenges no single element could handle alone.

Brass: The Golden Child of the Alloy Family

Take brass, for instance. This radiant alloy is born from the union of copper and zinc. Copper, known for its warmth and conductivity, lends its reddish glow and malleability to the mix. Zinc, the sturdy workhorse, contributes strength and resilience. Together, they create a material that's not only beautiful but also incredibly versatile.

Think of a trumpet gleaming under the stage lights, its brass body resonating with the musician's breath. Or picture intricate doorknobs and ornate fixtures, their golden sheen adding a touch of elegance to a room. Brass is even found in the gears and valves of machines, tirelessly working behind the scenes.

Stainless Steel: The Superhero of the Kitchen

Now, let's meet stainless steel, the champion of the culinary world. This alloy is like iron's cool, sophisticated cousin. Iron, strong but prone to rust, gets a makeover with the addition of chromium. This shiny element forms an invisible shield, protecting the iron from the ravages of water and oxygen.

Nickel joins the party, adding a touch of strength and flexibility, while carbon brings a dash of hardness to the mix. The result? A material that's not only incredibly durable but also resistant to stains and corrosion.

No wonder stainless steel is the star of our kitchens. Gleaming cutlery, sleek sinks, and sturdy cookware – they all owe their resilience to this remarkable alloy. But stainless steel's talents extend far beyond the kitchen. It's used in everything from skyscrapers and bridges to surgical instruments and spacecraft.

Why Alloys are Stronger Together

But how do these metallic mixtures become stronger than their individual components? It's all about disrupting the atomic order.

Imagine a perfectly organized parade of soldiers, all marching in sync. That's a pure metal, with its atoms lined up in neat rows. Now, imagine some mischievous kids joining the parade, disrupting the formation with their different sizes and unruly behavior. That's an alloy, with foreign atoms throwing the neat atomic arrangement into disarray.

This disruption might seem like chaos, but it's actually the secret to an alloy's strength. The foreign atoms act like tiny obstacles, making it harder for the atomic layers to slide past each other. It's like trying to walk through a crowded room – the more people in your way, the harder it is to move.

The Ever-Evolving World of Alloys

The world of alloys is constantly evolving, with scientists and engineers exploring new combinations and pushing the boundaries of material science. They're like alchemists of the modern age, creating materials with extraordinary properties.

Shape-memory alloys, for instance, can "remember" their original shape even after being bent or twisted. Imagine a medical implant that can be inserted into the body in a compact form and then expand to its intended shape once in place.

High-entropy alloys are another exciting frontier. These alloys are like a melting pot of elements, each contributing to a unique blend of properties. They're being explored for use in everything from jet engines to nuclear reactors.

Alloys: The Unsung Heroes of Modern Life

From the smartphones in our pockets to the cars we drive, alloys are the unsung heroes of modern life. They're the building blocks of our technological world, enabling advancements in everything from medicine and transportation to energy and communication.

So next time you encounter an object made of metal, take a moment to appreciate the ingenuity of alloys. These remarkable materials are a testament to the power of collaboration, proving that when metals come together, they can create something truly extraordinary.

Alloys: A Metal Mashup with Magical Properties

Imagine a world where metals were like solo artists, each with their own unique talents but limited range. Then, along came alloys, the supergroups of the material world! By blending different metals (and sometimes a dash of non-metal spice), we create materials with superpowers that their individual components could only dream of.

Why Alloys Rock:

Think of it like baking a cake. Flour alone might be bland, but add some sugar for sweetness, eggs for binding, and butter for richness, and boom – you've got a masterpiece! Alloys are similar. Let's break it down:

Strength and Toughness: Pure aluminum? Kind of wimpy. But mix in some magnesium, copper, and silicon, and suddenly you've got a material strong enough to build airplanes! Alloys can be engineered to be incredibly strong and resistant to bending or breaking.

Fighting Rust: Ever seen a rusty old car? That's what happens to iron when left to the elements. But add some chromium to the mix, and you get stainless steel – shiny, strong, and rust-proof! Perfect for everything from kitchen sinks to surgical tools.

Wear and Tear: Imagine your favorite shoes wearing out after just a few walks. Bummer, right? Alloys can be designed to resist wear and tear, making them ideal for things like drill bits, engine parts, and even artificial joints.

Electric and Magnetic Magic: Need a material that can handle high electrical currents or create a strong magnetic field? Alloys are your answer! They can be fine-tuned to have specific electrical and magnetic properties, making them essential for everything from toasters to MRI machines.

Temperature Tamers: Some alloys can handle extreme temperatures without expanding or contracting too much. This makes them perfect for delicate instruments, like watches and scientific equipment, that need to stay precise even when things heat up.

Alloy All-Stars:

Stainless Steel: The king of corrosion resistance, this alloy is a staple in kitchens, hospitals, and construction sites worldwide.

Brass: With its beautiful golden hue, brass is used in musical instruments, doorknobs, and even fancy light fixtures.

Bronze: Harder and more durable than pure copper, bronze has been used for centuries in everything from statues to ship propellers.

Aluminum Alloys: Lightweight yet strong, these alloys are essential for airplanes, cars, and even soda cans.

Titanium Alloys: These super-strong, lightweight alloys are used in everything from jet engines and spacecraft to high-performance bicycles and medical implants.

Peeking Inside Alloys:

Imagine looking at a LEGO creation. You can see how the different bricks fit together to create the final structure. Scientists can do something similar with alloys using powerful microscopes and diagrams. These tools help them understand how the arrangement of atoms affects the alloy's properties.

Alloy Adventures:

Jet Engines: The incredible heat and stress inside a jet engine require materials that can withstand extreme conditions. Nickel-based superalloys are up to the task, allowing planes to fly faster and farther.

Shape-Shifting Wonders: Shape memory alloys are like something out of a sci-fi movie! They can be bent and twisted, but when heated, they "remember" their original shape and spring back. This makes them perfect for medical implants like stents, which can be inserted into a blood vessel in a compact form and then expand to open up the blockage.

Extreme Environments: High-entropy alloys are a new breed of super-tough materials that can handle incredibly harsh conditions. Scientists are exploring their use in everything from nuclear reactors to deep-sea exploration.

The Future is Alloyed:

Alloys are constantly evolving, with researchers developing new combinations and techniques to create even more amazing materials. From lighter and stronger alloys for space travel to biocompatible alloys for medical implants, the possibilities are endless. So next time you use a fork, ride a bike, or fly in a plane, remember the magic of alloys – the unsung heroes of the material world!

Reactivity series

The Metal Mania: A Reactivity Riot!

Imagine a bustling city, filled with diverse metal citizens, each with their own unique personality and social life. Some are party animals, always ready to react and make a scene, while others prefer a quiet life, keeping to themselves. This is the world of metals, where their reactivity dictates their social status and how they interact with others.

The A-Listers: Potassium (K) and Sodium (Na)

These are the rockstars of the metal world, the life of every party! They're so reactive, they can't even chill with water without causing an explosive scene, releasing hydrogen gas like confetti. They're the ultimate divas, always ready to steal the show and kick out other metals from their compounds.

The Popular Kids: Calcium (Ca) and Magnesium (Mg)

These metals are still quite the partygoers, though not as wild as potassium and sodium. Calcium loves a good splash in water, creating a bubbly hydrogen show, while magnesium prefers a steamy sauna session. They're still quite the heartbreakers, capable of stealing partners from less reactive metals.

The Middle Ground: Aluminum (Al) and Zinc (Zn)

These metals are more laid-back, preferring a casual hangout with dilute acids rather than a wild water party. They're not as flashy as the A-listers, but they can still hold their own and displace less reactive metals from their compounds.

The Introverts: Iron (Fe) and Hydrogen (H)

Iron is the quiet one, content with reacting with dilute acids and forming subtle bonds. Hydrogen, though not a metal, is an honorary member of the group, acting as a benchmark for reactivity. Metals above hydrogen in the social hierarchy can easily steal its friends (electrons) away.

The Wallflowers: Copper (Cu), Silver (Ag), and Gold (Au)

These metals are the ultimate introverts, preferring to keep to themselves and avoid any dramatic reactions. They're content with their stable relationships and don't bother interacting with dilute acids or other metals.

The Drama of Displacement

The reactivity series is like a high school drama, filled with relationship swaps and social climbers. The more reactive metals, the popular kids, can easily steal partners (non-metal ions) from the less reactive ones. This creates a chain of displacement reactions, where metals higher up in the social ladder assert their dominance.

Corrosion: The Uninvited Guest

Corrosion is the villain of our story, attacking metals and causing them to deteriorate. Iron, unfortunately, is a prime target for this villain, forming rust when exposed to oxygen and water. But fear not! Zinc, the noble knight, comes to the rescue, sacrificing itself to protect iron from corrosion.

The End... or is it?

The reactivity series is a never-ending saga, filled with exciting twists and turns. By understanding this metal hierarchy, we can predict their behavior and use them wisely in various applications. So, keep exploring the fascinating world of metals and their reactive escapades!

Metals Gone Wild: A Reactivity Rodeo

Imagine the periodic table as a Wild West saloon. The metals are our rowdy cowboys, and their reactions with water and acids are the brawls that erupt! Some metals are like seasoned gunslingers, quick to draw and explode in a fiery reaction. Others are more like the town drunkard, slow to react and maybe just causing a bit of a fizzle.

Part A: Water Fight!

Potassium (K): This cowboy is a straight-up firecracker! Toss him in some water, and he'll explode with a "BOOM!" He's so eager to lose his electrons (like a poker player throwing money on the table), he reacts violently, creating potassium hydroxide and hydrogen gas. The heat is so intense it can even set the hydrogen on fire!

Sodium (Na): Sodium is like Potassium's slightly less crazy brother. He'll still cause a big splash in water, producing sodium hydroxide and hydrogen gas, but maybe won't set the whole saloon ablaze.

Calcium (Ca): Calcium is the cool-headed one of the bunches. He'll react with water, but it's more of a slow burn. He forms calcium hydroxide and hydrogen gas, but without the dramatic fireworks of his buddies.

The Trend: The further down the reactivity series we go, the calmer our cowboys get. Potassium is the wildest, Sodium is a bit tamer, and Calcium prefers a slow dance to a brawl.

Part B: Steamy Encounters

Magnesium (Mg): Magnesium is a bit picky. Cold water doesn't impress him; he needs the excitement of steam to get going. When things heat up, he reacts to form magnesium oxide and hydrogen gas.

Part C: Acid Antics

Now, let's move from the water through to the bar. Dilute hydrochloric acid is on tap, and our metals are ready for a different kind of showdown.

Magnesium (Mg): Magnesium jumps right in, reacting vigorously with the acid to produce magnesium chloride and hydrogen gas. This guy loves a good reaction, no matter what the liquid!

Zinc (Zn): Zinc is a bit more cautious. He'll react with the acid too, but it's a less intense affair. Zinc chloride and hydrogen gas are formed, but without the wild enthusiasm of magnesium.

Iron (Fe): Iron is the slow-talking, deliberate type. He takes his time reacting with the acid, eventually producing iron (II) chloride and hydrogen gas.

Copper (Cu), Silver (Ag), and Gold (Au): These guys are the real high rollers in the saloon. They're too cool for this acid stuff. They won't react with dilute hydrochloric acid at all.

The Trend: Again, reactivity rules the roost. Magnesium is the life of the party, Zinc is a bit more reserved, Iron takes it slow, and Copper, Silver, and Gold are too sophisticated for this ruckus.

The Moral of the Story:

The reactivity series is like a wanted poster in our Wild West saloon. It tells us who to watch out for and who's likely to cause the biggest commotion. The higher a metal is on the list, the more reactive it is, whether it's facing down water or acid.

Part 1: The Aluminum Mystery: Why Doesn't It Explode?

Imagine a superhero in disguise. That's aluminum! It's a metal so reactive it should be bursting into flames at the slightest touch, like a fireworks finale gone rogue. But hold on – your soda can isn't combusting in your hand. What's the secret?

Aluminum has a superpower: the ability to create an invisibility cloak. When exposed to air, it instantly reacts with oxygen, forming a super-thin, super-strong shield of aluminum oxide. This "oxide layer" is so tiny you can't even see it, but it's tougher than a knight's armor, protecting the aluminum from further reactions.

Think of it like this: you're trying to get to the juicy aluminum inside, but this fierce bodyguard, the oxide layer, is blocking your way. It's waterproof, clingy (it never let's go of aluminum!), and stubbornly unreactive.

This explains why aluminum, despite its explosive potential, is happy to chill in your kitchen as pots and pans, or soar through the sky as airplanes. It's all thanks to its incredible invisibility cloak!

But what if we want to see aluminum's true reactive nature? Well, we need to get past the bodyguard. We can try to scratch it off, melt it with acid, or even use a trick with mercury to disrupt it. Once the shield is down, get ready for some fireworks!

Aluminum in Action:

Aircraft: Aluminum's lightness and strength (thanks to the oxide layer!) make it perfect for airplanes. Imagine if planes were made of a metal that rusted easily – yikes!
Food packaging: Ever noticed how your aluminum foil doesn't react with your food? Thank the oxide layer for keeping your snacks safe and tasty.
Construction: Aluminums used in buildings because it's strong, light, and resists the elements. Again, the oxide layer is the unsung hero, protecting against rain, wind, and sun.

Part 2: The Metal Olympics: Who's the Most Reactive?

Imagine a metal Olympics, where metals compete to see who's the most reactive. How do we figure out who wins the gold medal?

We set up some challenges! We make them react with water and acid, like a crazy obstacle course. Some metals will go wild, fizzing and bubbling like a shaken soda bottle. Others might just sit there, unimpressed.

Here's how we figure out the winners:

Speed demons: The faster a metal reacts, the more reactive it is. Think of it like a sprint to the finish line!
Heat wave: Some reactions release a lot of heat. The hotter it gets, the more reactive the metal. Imagine a fiery explosion versus a lukewarm bath.

Transformation: The type of product formed also matters. Some metals create more stable compounds, showing their superior reactivity. It's like transforming into a powerful superhero!

Let's look at an example:

Imagine Metal A reacts with water like crazy, producing tons of bubbles (hydrogen gas). But Metal B just sits there, doing nothing. Clearly, Metal A is the reactivity champion!

Why does this matter?

The "reactivity series" is like a metal popularity contest, showing us who's the most reactive. This helps us:

Predict reactions: We can guess what will happen when different metals meet.
Choose the right metal: We can pick the best metal for a job. You wouldn't want a super reactive metal for your jewelry, would you?
Understand corrosion: Some metals rust easily because they're high on the reactivity list. It's like they're eager to react with anything!
This "metal Olympics" helps us understand the amazing world of metals and how they interact with the world around them.

Rust: The Silent Thief That Eats Metal

Imagine a tiny, invisible army of gremlins gnawing away at your prized possessions. That's kind of what rust is like. This reddish-brown menace, scientifically known as hydrated iron (III) oxide, is a metal's worst nightmare. It's not just a cosmetic issue; rust weakens structures, causing everything from bridges to bicycles to crumble.

The Rust Recipe: More Than Just Oxygen

You might think rust is just iron reacting with oxygen, but it's a bit more complex than that. It's like baking a cake – you need the right ingredients in the right conditions.

Iron (Fe): The main course, the star of the show. Whether it's pure iron or an alloy like steel, it's all susceptible to the gremlins' attack.
Oxygen (O_2): The hungry guest, always eager to snatch electrons from iron atoms. It's everywhere, lurking in the air and dissolved in water.
Water (H_2O): The mischievous accomplice, helping the oxygen and iron get together. It acts like a highway for the tiny charged particles that drive the rusting process.
Electrolyte: Think of this as the secret ingredient that speeds things up. Dissolved salts, acids, or other impurities in the water are like adding turbochargers to the gremlins.

The Rusting Tango: A Chemical Dance

Rusting is a two-step process, a delicate dance between oxidation and reduction.

Oxidation: Iron atoms lose electrons, like shedding a coat, and become iron (II) ions (Fe^{2+}).
Reduction: Oxygen molecules gain those electrons, like putting on a warm sweater, and become hydroxide ions (OH^-).
These ions then mingle and react, eventually forming the dreaded rust.

Factors Influencing the Rusting Rate

Just like some cakes bake faster than others, several factors influence how quickly rust forms:

pH: Acidic environments are like turning up the oven, while alkaline conditions slow things down.
Temperature: Higher temperatures make the gremlins work faster.
Salt Concentration: More salt is like adding more fuel to the fire.
Metal Purity: Pure iron is like a tough cookie, resisting the gremlins better than impure iron or alloys.
Surface Area: The more surface area exposed, the more room the gremlins have to feast.
Mechanical Stress: Stressed metal is like a cracked cookie, easier for the gremlins to break into.

Barrier Methods: Shielding Metal from the Elements

To protect our precious metal possessions, we need to build barriers, like tiny fortresses, to keep the rust gremlins at bay.

Painting: A classic defense, like giving the metal a coat of armor. Different paints offer varying levels of protection, so choose wisely.
Greasing: Like slathering the metal in slippery oil, making it hard for the gremlins to get a grip. Perfect for moving parts.
Coating with Plastic: A more modern approach, encasing the metal in a durable plastic shield.
Other barrier methods include galvanizing (sacrificing a layer of zinc to protect the iron), tin plating, and electroplating.

Rust in the Real World

The Golden Gate Bridge: Battling the elements with a special rust-inhibiting paint.
Cars: Protected by a combination of paint, galvanizing, and undercoating.
Pipelines: Shielded from underground and underwater corrosion with specialized coatings.

The Takeaway

Rust is a formidable foe, but by understanding its nature and employing effective barrier methods, we can keep our metal treasures safe from its clutches.

Imagine a Tiny Superhero...

...so small it could hang onto a metal beam with its microscopic hands. This superhero, made entirely of zinc, spends its days battling the evil villain, Rust, who's always trying to gobble up bridges, ships, and even your backyard swing set.

Our zinc superhero is a "sacrificial anode," a real-life hero in the world of corrosion prevention. It works like this: metals have different levels of "eagerness" to give up their electrons. Think of it like a game of hot potato – some metals can't wait to toss that electron away. Zinc is one of those eager metals.

When zinc is near a less eager metal, like iron (what your swing set is made of), it shouts, "I'll take that hot potato!" and grabs the electrons. Rust, who loves to steal electrons from iron and turn it into a flaky mess, gets completely blocked. Our zinc superhero sacrifices itself, slowly corroding away, but in doing so, it saves the iron from a rusty fate.

This is sacrificial protection in action! It's like a bodyguard for metal, taking the hits so the important stuff stays safe.

Galvanizing: The Superhero's Cape

Now, imagine our zinc superhero puts on a special cape – a shiny coat of zinc called "galvanizing." This cape has two amazing powers:

Invisibility Shield: The cape hides the iron from Rust, making it practically invisible to the villain.
Super Strength: Even if Rust manages to poke a hole in the cape, our zinc superhero is still there to fight! It throws itself into the fray, sacrificing itself to protect the iron underneath.

This is why galvanized steel is so awesome. It's like having a double layer of protection against rust – a secret weapon for everything from skyscrapers to guardrails.

Real-Life Superhero Stories:

The Golden Gate Bridge: This iconic bridge is like a giant, red superhero, and its secret weapon is galvanizing. The zinc coating keeps it safe from the salty sea air and strong winds, allowing it to stand tall and proud for almost a century.
Your Car: Peek under your car, and you might see some shiny parts. That's galvanized steel, protecting your car's underbelly from rusting away.

The Future of Superheroes

Scientists are always working on new ways to improve our zinc superhero. They're creating super-strong alloys and even giving it a jetpack with "thermal spraying" technology. This means our tiny hero can fly onto huge structures like ships and bridges, protecting them from Rust's evil clutches.

So next time you see a shiny piece of metal, remember our tiny zinc superhero, bravely fighting to keep our world safe from rust, one electron at a time.

Extraction of metals

The Metal Kingdom: A Reactivity Saga

Imagine a grand kingdom of metals, each vying for the throne of "Easiest to Extract." Their nobility? Their reactivity – a measure of their eagerness to ditch their electrons and form alliances with other elements. The more reactive, the higher their status in this royal court.

The Highborn:

At the top, we have the impulsive royals like Potassium, Sodium, Calcium, Magnesium, and Aluminum. Always eager to mingle, they're never found alone in nature. They're bound to other elements in ores, like nobles clinging to their court. Extracting them is like convincing a popular monarch to abdicate – it requires a powerful force, like the electrifying jolt of electrolysis.

The Middle Class:

Further down the line, we have the more level-headed Zinc, Iron, Lead, and Copper. They're still reactive, but less prone to drama. These metals can be coaxed from their ores with a bit of persuasion, like a good old-fashioned chat with Carbon. This "carbon reduction" is like a negotiation, where carbon convinces the metal to leave its compound and join forces with it instead.

The Reclusive:

At the bottom dwell the introverted nobles, Silver and Gold. Content in their own company, they're often found in their pure, elemental form. Extracting them is less about persuasion and more about discovery. Think of panning for gold – it's like finding a shy hermit in their secluded dwelling.

The Aluminum Affair: A Tale of Electrolysis

Aluminum, a high-ranking noble, is notoriously difficult to extract from its ore, bauxite. It's like trying to separate a social butterfly from a bustling party. Enter the ingenious Hall process, a method as dramatic as a royal ball.

Cryolite: The Master of Ceremonies: Cryolite, a salt, is like the master of ceremonies, ensuring the smooth flow of the event. It lowers the melting point of the aluminum ore, making it more manageable, like calming the excited partygoers. It also boosts the conductivity of the molten mix, allowing the electric current to flow freely, like ensuring the music reaches every corner of the ballroom.

Carbon Anodes: The Sacrificial Dancers: The carbon anodes are like sacrificial dancers, consumed in the process. Oxygen, produced at the anode, reacts with the carbon, forming carbon dioxide, like a fiery dance that leaves the dancers exhausted. These anodes need constant replacement, a costly affair, much like the expenses of maintaining a grand ballroom.

Electrolysis: The Grand Finale: The electrolysis itself is the grand finale. At the cathode, aluminum ions are drawn in, gaining electrons and transforming into molten aluminum, like guests receiving gifts from the host. At the anode, oxide ions lose electrons, forming oxygen gas, like partygoers shedding their inhibitions.

The Aftermath:

The molten aluminum, now free from its ore, is collected and cast into ingots, like honored guests departing with precious souvenirs. The process, though energy-intensive, is a testament to human ingenuity, like a perfectly orchestrated royal ball that leaves everyone in awe.

Beyond the Ballroom:

The story of aluminum doesn't end there. Scientists are constantly seeking ways to make the process more sustainable, like finding ways to reuse the ballroom decorations. Recycling aluminum requires a fraction of the energy needed for primary production, a noble effort in preserving our resources.

In Conclusion:

The extraction of metals is a fascinating tale of reactivity, ingenuity, and sustainability. From the electrifying drama of electrolysis to the quiet elegance of physical extraction, each method reflects the unique character of the metals involved. And as we continue to explore and innovate, we ensure that the Metal Kingdom remains a source of wonder and prosperity for generations to come.

Imagine a Giant, Fiery Beast

Forget those dry textbook descriptions! Picture a colossal, fire-breathing dragon, its belly glowing with unimaginable heat. That's our blast furnace, a monstrous contraption built to wrestle iron from the clutches of its ore, hematite.

1. The Dragon's Breath: Coke Combustion

Deep within this beast's belly, we unleash a fiery inferno by burning coke, a coal-derived fuel packed with carbon. This isn't just any bonfire; it's a raging inferno that would make a blacksmith blush.

Heatwave: This intense heat transforms the furnace into a crucible, reaching temperatures hotter than molten lava. It's like the dragon's fiery breath, melting everything in its path.
CO2 Creation: The burning coke also releases carbon dioxide, a gas that's about to play a surprising role.

2. A Gaseous Transformation: CO2 to CO

As the carbon dioxide rises through the furnace, it's like walking into a sauna...of white-hot coke! This intense heat forces the CO2 to change its identity, becoming carbon monoxide (CO). Think of it as the dragon's exhaled breath, now ready to perform some serious magic.

3. Breaking the Chains: Freeing the Iron

Now, imagine our hematite ore (iron oxide) tumbling down from the top, like hapless knights facing the dragon. But instead of flames, they meet the rising carbon monoxide. This clever gas steals the oxygen from the iron oxide, setting the pure, molten iron free! It's a chemical heist, leaving the iron to flow like liquid gold to the bottom of the furnace.

4. The Limestone Purifier: Dealing with the 'Riff-Raff'

Iron ore isn't always pure; it often comes with unwanted baggage – impurities like silica. To get rid of these pesky hangers-on, we introduce limestone, a rock that acts like a bouncer at the dragon's lair. The intense heat breaks down the limestone, creating calcium oxide, our cleaning agent.

5. Slag: The Protective Shield

This calcium oxide then teams up with the silica impurities to form slag, a molten concoction that floats on top of the liquid iron. Think of it as a protective shield, preventing the precious iron from getting contaminated or re-oxidized. The slag is then drained off, like removing the unwanted guests from the dragon's party.

The Iron Symphony

Extracting iron from hematite is like conducting a fiery orchestra, with each chemical reaction playing its part. It's a process that has fueled civilizations for centuries, and it all happens within this magnificent, fiery beast – the blast furnace!

Bonus Beats:

This dragon never sleeps! The blast furnace runs 24/7, with raw materials constantly fed in and molten iron tapped out.
The furnace is a master of recycling, reusing the hot gases to preheat the incoming air, saving energy and keeping the dragon's breath burning bright.
The iron produced is just the beginning. It's further refined into steel, the backbone of our modern world.

Chemistry of the environment

Water

Water: The Sneaky Little Molecule and How We Catch It

Water, water everywhere! It's the lifeblood of our planet, essential for every creature and plant. But sometimes, we need to know if water is lurking where it shouldn't be, or if the water we have is truly pure. That's where a bit of chemistry magic comes in.

The Color-Changing Tricksters

Imagine substances that change color like chameleons when they sense water. That's exactly what anhydrous cobalt (II) chloride and anhydrous copper (II) sulfate do!

Cobalt (II) Chloride: The Dramatic Blue-to-Pink Transformation

Think of anhydrous cobalt (II) chloride as a grumpy blue fellow. He's perfectly content being blue, but the moment he feels a drop of water, he blushes a rosy pink. This dramatic color change is a surefire sign that water is present. Scientists use this trick to detect leaks, measure humidity, and even create simple weather indicators.

Copper (II) Sulfate: From Wallflower White to Bold Blue

Anhydrous copper (II) sulfate is a bit more subtle. It starts as a plain white powder, but add a touch of water, and it transforms into a vibrant blue crystal. This "blue vitriol," as it's known, is a telltale sign of water's presence. Chemists use this to test for water in organic solvents and even to check if a salt has lost its water content.

The Temperature Test: Is Your Water a Boiling Point Bully?

Pure water has a very specific boiling point – 100 °C (212 °F). But if there are impurities lurking in the water, they can raise the boiling point. Think of it as the impurities being bullies, making the water work harder to reach its boiling point. By carefully measuring the boiling point, scientists can tell if water is pure or if it's harboring unwanted guests.

Melting Point Mysteries

Similarly, pure water freezes at 0 °C (32 °F). But impurities can disrupt the water molecules and lower the freezing point. It's like the impurities are tripping up the water molecules as they try to form ice. By checking the melting point, we can get another clue about the purity of our water.

The Limitations of Our Water Detectives

While these tests are clever, they do have their limits. The color-changing tests can't tell us exactly how much water is present, and they might be fooled by other substances that react similarly. The temperature tests are also not very sensitive to tiny amounts of impurities.

The Importance of Water Purity

Knowing if water is pure is important in many areas of life, from ensuring safe drinking water to conducting precise scientific experiments. By using these clever chemical and physical tests, we can unveil the secrets hidden within this seemingly simple substance.

Imagine a world where chemists are like chefs, concocting magical potions instead of delicious dishes. Just as a chef wouldn't use muddy water to make a delicate soup, a chemist wouldn't use tap water for sensitive experiments.

That's where distilled water comes in – it's the secret ingredient that ensures the magic works as expected. Think of it as the purest form of water, stripped bare of all the extras that tap water carries. No minerals, no gases, no tiny critters – just pure H2O.

Why is this so important? Well, those "extras" in tap water can be quite the troublemakers in the lab. They can mess with reactions, throw off measurements, and even contaminate delicate equipment. It's like adding a pinch of salt to a cake recipe – a tiny change can make a big difference!

Distilled water, on the other hand, is like a blank canvas. It lets the chemist be the artist, controlling every element of the experiment. It's the silent partner that ensures accurate results, prevents unexpected surprises, and keeps the lab running smoothly.

So, next time you see a bottle of distilled water, remember its unsung role in the world of chemistry. It's the quiet achiever, the behind-the-scenes hero that allows science to flourish.

Here are a few real-world examples of how distilled water works its magic:

Titration: Imagine trying to measure the exact amount of acid in a solution using a burette. Any leftover impurities in the burette could throw off the measurement, like a wonky measuring cup. Distilled water rinses away those impurities, ensuring a precise measurement.
Spectrophotometry: This technique uses light to analyze substances. Distilled water provides a clean baseline, like a perfectly clear window, allowing scientists to see the true colors of their samples.
Electrochemistry: In this field, scientists study the relationship between electricity and chemical reactions. Distilled water prevents unwanted reactions, like a bodyguard protecting a VIP.

Beyond the lab, distilled water also plays a part in our everyday lives:

Cars: It keeps car batteries and cooling systems running smoothly, preventing corrosion and mineral buildup.
Homes: It prevents mineral deposits in steam irons and humidifiers, extending their lifespan.
Aquariums: It helps maintain a healthy environment for fish, preventing the buildup of harmful minerals.
In conclusion, distilled water is a bit like a superhero in disguise. It might seem ordinary, but it plays a crucial role in science, industry, and even our homes. So, let's raise a glass (of distilled water, of course!) to this unsung hero of the chemical world!

The Secret Life of Water: A Look at What Lurks Beneath the Surface

Water. It's the essence of life, a shimmering elixir that quenches our thirst and sustains our world. But don't let its crystal-clear facade fool you. Water, especially from natural sources, is a bustling metropolis of unseen residents and dissolved secrets. Think of it as a microscopic Grand Central Station, with molecules and particles constantly coming and going, leaving their mark on this precious liquid.

Let's dive in and meet some of the characters that call water home:

(a) The Breath of Life: Dissolved Oxygen

Imagine water as a vast lung, inhaling oxygen from the air above. This oxygen is the lifeblood of aquatic creatures, from the tiniest plankton to the mightiest fish. But like a delicate dance, the amount of oxygen in the water depends on a variety of factors.

Temperature: Picture a cold mountain stream, its waters invigorated with oxygen. Now imagine a warm, stagnant pond, struggling to breathe. Cold water, like a tightly sealed container, can hold onto more oxygen than its warmer counterpart.
Turbulence: Think of a waterfall, its cascading waters churning and mixing with the air, eagerly absorbing oxygen. Turbulence is like a vigorous shake of a soda bottle, infusing the water with a refreshing dose of oxygen.
Organic Matter: But beware the lurking villains! Decaying leaves, sewage, and other organic matter are like oxygen thieves, consuming it as they decompose. Too much of this, and the water becomes a suffocating trap for its inhabitants.

Examples:

A Thriving Community: A clear, rushing stream, teeming with fish and insects, is a testament to the power of dissolved oxygen.
A Silent Struggle: A murky pond, choked with algae and devoid of life, reveals the devastating consequences of oxygen depletion.
The Gulf's Silent Scream: The "dead zone" in the Gulf of Mexico is a chilling example of what happens when excess nutrients trigger an

oxygen-consuming algal bloom, leaving a vast underwater graveyard in its wake.

(b) The Metallic Medley: Metal Compounds

Water, in its journey through the earth, picks up souvenirs along the way. These souvenirs come in the form of metal compounds, leached from rocks and soils. Some, like calcium and magnesium, are welcome guests, contributing to water's "hardness" and even providing essential nutrients. But others, like lead and mercury, are unwelcome intruders, posing serious health risks.

Geological Influences: The type of rocks and soil that water encounters on its journey shapes its metallic profile.
Human Impact: Mining, industrial waste, and urban runoff can introduce a toxic cocktail of heavy metals into our water supplies.

Examples:

The Scale of Things: Hard water, rich in calcium and magnesium, might leave unsightly deposits on your faucets, but it's generally harmless to drink.
The Acidic Assault: Acid mine drainage, like a corrosive acid, leaches heavy metals from abandoned mines, poisoning streams and rivers.
The Flint Water Crisis: This tragic event serves as a stark reminder of the devastating consequences of lead contamination in our drinking water, leaving a legacy of health problems in its wake.

(c) The Uninvited Guest: Plastics

Plastic, the ubiquitous material of our modern world, has infiltrated even the most pristine waters. Microplastics, tiny fragments of plastic debris, are now found in every corner of the globe, from the depths of the ocean to the water we drink.

A Persistent Problem: Plastics are the unwelcome houseguests that overstay their welcome, lingering in the environment for centuries.
A Threat to Life: Marine animals often mistake microplastics for food, leading to starvation, entanglement, and internal injuries.

Examples:

A Fatal Meal: Seabirds, deceived by the appearance of plastic fragments, ingest them, filling their stomachs with indigestible trash.
Hidden Dangers: Microplastics have infiltrated our food chain, lurking in seafood, table salt, and even drinking water.
The Great Pacific Garbage Patch: This swirling vortex of plastic debris in the Pacific Ocean is a monument to our throwaway culture.

(d) The Unwanted Waste: Sewage

Sewage, the byproduct of our daily lives, carries a hidden burden of organic matter, nutrients, pathogens, and chemicals. When untreated or improperly treated, it contaminates our water sources, posing a serious threat to human and environmental health.

Disease and Depletion: Sewage pollution can spread waterborne diseases, deplete oxygen levels, and introduce harmful substances into the food chain.
The Importance of Treatment: Proper sewage treatment is essential for safeguarding our water resources and protecting public health.

Examples:

Invisible Enemies: Cholera and typhoid, deadly waterborne diseases, lurk in contaminated water, waiting to strike.
Choking the Life Out of Our Waters: Sewage discharge can fuel algal blooms, creating oxygen-deprived dead zones in rivers and lakes.
The Ganges River: This sacred river, revered by millions, is tragically burdened by sewage and industrial waste, posing a significant health risk to those who depend on it.

(e) The Microscopic Menaces: Harmful Microbes

Water teems with microscopic life, a hidden world of bacteria, viruses, and protozoa. While most are harmless, some can cause debilitating diseases in humans and animals.

Sources of Contamination: Poor sanitation, animal waste, and inadequate water treatment can all contribute to the presence of harmful microbes in our water.

Examples:

E. coli: This bacterial villain, a common resident of our intestines, signals fecal contamination when found in water.
Cryptosporidium: This microscopic parasite, resistant to chlorine disinfection, can cause widespread outbreaks of gastrointestinal illness.
The Milwaukee Outbreak: The 1993 Cryptosporidium outbreak in Milwaukee, which sickened over 400,000 people, highlights the vulnerability of our water supplies to microbial contamination.
(f) & (g) The Nutrient Overload: Nitrates and Phosphates

Nitrates and phosphates, essential nutrients for plant growth, can become unwelcome pollutants when they enter our water in excess. Fertilizers and detergents are the main culprits, contributing to a nutrient overload that disrupts aquatic ecosystems.

Eutrophication: Excess nutrients fuel algal blooms, which choke waterways, deplete oxygen, and harm aquatic life.

Examples:

Agricultural Runoff: Fertilizer-laden runoff from fields contaminates streams and rivers, contributing to the growth of harmful algal blooms.
Blue Baby Syndrome: High nitrate levels in drinking water can cause methemoglobinemia in infants, a condition that reduces the blood's ability to carry oxygen.
Lake Erie's Struggles: Recurring algal blooms in Lake Erie, fueled by excess nutrients, threaten drinking water supplies and disrupt the lake's delicate ecosystem.

Conclusion:

Water, the elixir of life, is a complex and vulnerable resource. By understanding the hidden world within it, we can appreciate its fragility and work to protect it from the threats it faces. From emerging contaminants to the impacts of climate change, the challenges are many, but so are the solutions. By embracing sustainable water management practices and staying informed about the latest research, we can ensure that clean, safe water remains a birthright for all.

Imagine a world where water, the elixir of life, is tainted with invisible foes. Heavy metals like lead, mercury, arsenic, and cadmium lurk in the shadows, threatening to poison our bodies and minds. Microplastics, those tiny plastic invaders, infiltrate our waterways, wreaking havoc on aquatic life and potentially harming us. Sewage and harmful microbes transform our pristine waters into breeding grounds for disease. And excessive nitrates and phosphates, like overgrown weeds, choke the life out of our aquatic ecosystems.

But fear not, for we have the power to fight back! Like alchemists of old, we've developed ingenious methods to purify our precious water. We use sedimentation and filtration to remove visible impurities, like sifting dirt from gold. Activated carbon, like a molecular sponge, traps unwanted tastes and odors, leaving the water fresh and pure. And chlorination, our secret weapon, vanquishes harmful microbes, ensuring our water is safe to drink.

Think of the Walkerton E. coli outbreak in Canada, a stark reminder of the importance of vigilant water treatment. Or the arsenic contamination in Bangladesh, a silent killer lurking in the groundwater. And let's not forget the microplastic pollution in the Great Lakes, a growing threat to our freshwater ecosystems.

These stories are not just tales of caution, but calls to action. We must continue to research and invest in water treatment technologies, like developing new weapons in the fight for clean water. We must protect our water sources from pollution, like guardians of a sacred treasure. And we must educate ourselves and others about the importance of water conservation, like spreading the gospel of a precious resource.

Together, we can ensure that clean water is not a luxury, but a birthright for all. We can transform our relationship with water from one of careless consumption to one of reverence and respect. And we can create a world where water, the lifeblood of our planet, flows pure and free for generations to come.

Fertilizers

Ammonium Salts and Nitrates: The Unsung Heroes of Plant Growth

In the bustling world of plant nutrition, nitrogen takes center stage. It's the key ingredient in chlorophyll, the pigment that gives plants their vibrant green color and allows them to harness the sun's energy. Nitrogen is also a building block for amino acids, the workhorses of the plant kingdom, responsible for everything from growth and repair to defense against pests and diseases.

But where do plants get this vital nutrient? Enter ammonium salts and nitrates, the dynamic duo of nitrogen fertilizers.

Ammonium Salts: The Slow and Steady Providers

Ammonium salts, like the reliable ammonium sulfate and ammonium nitrate, are the tortoises of the nitrogen world. They release nitrogen gradually, providing a steady supply to plants over time. This slow-release action is perfect for colder climates or soils rich in organic matter, where microbial life takes its time to convert ammonium to nitrates, the form of nitrogen most readily absorbed by plants.

Nitrates: The Speedy Nitrogen Boosters

Nitrates, on the other hand, are the hares of the nitrogen race. Calcium nitrate and potassium nitrate, for example, are readily absorbed by plants, delivering a quick nitrogen boost. This rapid action makes them ideal for crops with high nitrogen demands or those needing a quick pick-me-up during critical growth stages.

NPK Fertilizers: The Balanced Diet for Plants

While nitrogen is essential, it's not the only nutrient plants need to thrive. Phosphorus and potassium are also crucial, playing vital roles in root development, energy transfer, water regulation, stress resistance, and enzyme activation.

NPK fertilizers offer a balanced blend of these three essential nutrients, ensuring plants receive a well-rounded diet. The NPK ratio, indicated by three numbers on fertilizer labels, represents the percentage of nitrogen, phosphorus (expressed as phosphate), and potassium (expressed as potash) by weight.

Choosing the Right NPK Fertilizer

Selecting the right NPK fertilizer is like choosing the right diet for your plants. It depends on factors like the type of crop, soil conditions, and growth stage. Leafy greens crave nitrogen, while fruiting crops need more phosphorus and potassium. Soil tests can help identify nutrient deficiencies, guiding you toward the perfect NPK match.

Beyond NPK: The Micronutrient Support System

While NPK fertilizers provide the main course, micronutrients are the essential vitamins and minerals for plant health. Iron, zinc, manganese, and copper, though needed in smaller amounts, are crucial for various physiological processes. Deficiencies in these micronutrients can lead to stunted growth, reduced yields, and poor quality.

Emerging Trends in Fertilizer Technology

The world of fertilizers is constantly evolving, with scientists and researchers developing new and innovative ways to improve nutrient use efficiency, reduce environmental impact, and enhance crop productivity.

Controlled-Release Fertilizers: These fertilizers are like time-release capsules, gradually releasing nutrients over time, minimizing losses and reducing the need for frequent applications.
Bio stimulants: These products are like plant probiotics, containing natural substances like seaweed extracts and beneficial microbes that stimulate growth and enhance nutrient uptake.

Precision Agriculture: This technology is like GPS for your plants, using sensors and data analytics to tailor nutrient delivery to specific crop needs and soil conditions.

Conclusion

Ammonium salts and nitrates are the unsung heroes of plant growth, providing the essential nitrogen that fuels life. NPK fertilizers offer a balanced diet, ensuring plants receive all the macronutrients they need to thrive. By understanding the roles of these fertilizers and the factors influencing their effectiveness, we can cultivate healthier, more productive crops while minimizing our environmental impact. As technology advances, we can look forward to even more innovative and sustainable approaches to plant nutrition, ensuring a bountiful harvest for generations to come.

The Air We Breathe: A Tale of Two Atmospheres
Imagine our atmosphere as a grand, swirling cocktail. In its purest form, it's a refreshing mix, mostly nitrogen with a healthy shot of oxygen, and a dash of other ingredients like argon and a sprinkle of carbon dioxide. This is the "clean air" cocktail – invigorating and essential for life.

But here's where the story takes a twist. Human activity has become a rowdy bartender, carelessly tossing in extra shots and strange concoctions, muddling the once pristine drink. Let's take a closer look at some of the unwelcome additions to our atmospheric cocktail:

The Usual Suspects:

Carbon Dioxide (CO2): Like an overly generous pour of soda, excessive CO2 is making our atmosphere fizzy in a way that's not so refreshing. Cars, power plants, and factories are the biggest culprits, guzzling fossil fuels and exhaling CO2. It's like they're trying to win a belching contest, with our planet as the unwilling judge.
Carbon Monoxide (CO) and Particulates: These are the unwanted dregs – the smoky residue from fires, car exhaust, and industrial processes. Imagine a smoky bar where you can barely see across the room. That's what these pollutants do to our air.
Methane (CH4): Think of methane as the unexpected guest who overstays their welcome. It seeps from landfills, escapes from gas pipelines, and even emerges from the, ahem, digestive processes of our bovine friends.
Oxides of Nitrogen (NOx): These are the hot-headed troublemakers, formed in the fiery furnace of car engines and power plants. They react with other ingredients in the atmosphere to create a hazy smog, like a brawl breaking out in our precious cocktail.
Sulfur Dioxide (SO2): This pungent addition comes courtesy of burning fossil fuels, especially coal. It's like someone spilled a bottle of vinegar in our drink, making our eyes water and our lungs burn. Remember the Great Smog of London? It was a tragic cocktail party gone wrong, where industrial emissions and stagnant weather conditions combined to create a deadly fog. It's a stark reminder of what happens when we let pollution spiral out of control.

And what about Beijing's air quality? It's like the city is constantly shrouded in a hazy hangover, a consequence of rapid industrial growth and heavy reliance on coal.

The good news is that we can change the recipe. By switching to cleaner energy sources, tightening our belts on emissions, and embracing sustainable practices, we can restore our atmosphere to its former glory. It's time to ditch the polluting ingredients and create a cocktail that's healthy, vibrant, and worthy of a toast.

Adverse Effects of Air Pollutants: A Deep Dive (2024 Update)

The Invisible Villains in Our Air: A Tale of Six Pollutants

In the bustling heart of our cities and the quiet corners of our countryside, an invisible battle rage. It's a battle between the air we breathe and the pollutants that seek to poison it. These pollutants, like malevolent spirits, come in many forms, each with its own unique way of wreaking havoc on our health and our planet.

1. Carbon Dioxide (CO_2): The Silent Saboteur

CO_2, the invisible gas we exhale with every breath, has a dark side. While essential for life, its overabundance in the atmosphere is like a fever for our planet, causing a cascade of devastating effects.

Global Warming and Climate Change: CO_2 acts like a blanket, trapping heat and causing our planet to warm. This warming leads to extreme weather events, like the scorching 2023 European heatwave that left a trail of destruction and despair.
Sea Level Rise: As glaciers melt and oceans expand, coastal communities face the threat of being swallowed by the rising tides. The Maldives, a paradise on Earth, is fighting a desperate battle against this watery invasion.
Ocean Acidification: The ocean, our planet's lifeblood, is becoming increasingly acidic as it absorbs excess CO_2. This acidity is like a poison, slowly killing marine life, including the vibrant coral reefs that are home to countless creatures.

2. Carbon Monoxide (CO): The Silent Assassin

CO, a colorless, odorless gas, is a stealthy killer. It silently replaces oxygen in our blood, depriving our organs of the life-giving element.

Health Impacts: CO poisoning can cause headaches, dizziness, and even death. It's like a thief in the night, stealing our breath and leaving us gasping for air.
Sources and Vulnerable Populations: Faulty heating systems, gas appliances, and vehicle exhaust are all sources of CO. The young, the elderly, and those with respiratory problems are particularly vulnerable to its deadly embrace.

3. Particulate Matter (PM): The Tiny Terror

PM, a mixture of solid particles and liquid droplets, is like a swarm of microscopic invaders. These tiny particles can penetrate deep into our lungs, causing inflammation and disease.

Health Impacts: PM2.5, the smallest and most dangerous type of particulate matter, can cause respiratory problems, cardiovascular disease, and even lung cancer. It's like a slow poison, gradually eroding our health.
Sources and Vulnerable Populations: PM2.5 comes from various sources, including vehicle exhaust, industrial emissions, and wildfires. Children, the elderly, and those with existing health conditions are most at risk.

4. Methane (CH4): The Potent Pollutant

Methane, a powerful greenhouse gas, is like a fiery dragon, trapping heat and fueling global warming.

Climate Change Impacts: Methane is a major contributor to climate change, causing temperatures to rise and weather patterns to become more erratic. It's like a raging inferno, threatening to consume our planet.
Sources: Methane comes from various sources, including natural gas leaks, livestock, and landfills. It's a reminder that even our everyday activities can have a significant impact on the environment.

5. Oxides of Nitrogen (NOx): The Reactive Rebels

NOx, a group of reactive gases, are like mischievous imps, causing trouble wherever they go.

Acid Rain: NOx contributes to acid rain, which damages forests, lakes, and buildings. It's like a corrosive acid, eating away at our natural and man-made structures.
Photochemical Smog: NOx also contributes to smog, a thick haze that blankets our cities and harms our health. It's like a suffocating blanket, making it difficult to breathe and see.

6. Sulfur Dioxide (SO2): The Pungent Poison

SO2, a colorless gas with a pungent odor, is like a noxious fume, irritating our lungs and contributing to acid rain.

Acid Rain: SO2 is a major contributor to acid rain, with all its associated environmental and health problems. It's like a corrosive rain, damaging our ecosystems and infrastructure.
Respiratory Problems: SO2 can also cause respiratory problems, especially in those with pre-existing conditions. It's like an irritant, making it difficult to breathe and causing discomfort.

The Battle for Clean Air

The fight against air pollution is a battle for our health, our planet, and our future. It's a battle we can't afford to lose. By understanding the adverse effects of these invisible villains, we can take steps to reduce our emissions, protect our health, and create a cleaner, healthier world for ourselves and future generations.

Remember: The air we breathe is a shared resource. Let's work together to protect it and ensure that everyone has access to clean, healthy air.

Imagine our planet as a cozy greenhouse, bathed in warm sunlight. The glass panes, like our atmosphere, let the sun's rays in, keeping us snug and comfortable. But lately, we've been adding extra layers of glass – that's the CO2 and methane we pump into the air.

The more layers we add, the hotter it gets inside. This is the greenhouse effect in action, and it's turning our comfortable planet into a sweltering hothouse.

Think of CO2 as a thick blanket, and methane as a super-charged heat-trapping sheet. Both trap the sun's warmth, but methane is like that extra fluffy blanket you pull out on the coldest nights – it holds in way more heat. The problem is, we're piling on both blankets at an alarming rate, and the planet is starting to sweat.

So, what can we do? Imagine planting trees as tiny superheroes, each one a miniature air purifier, sucking CO2 out of the air with every breath. Picture a world where we eat less meat, easing the burden on our planet and reducing methane emissions from livestock. Envision a future powered by the sun and wind, where fossil fuels are relics of the past, and hydrogen, a clean energy superhero, fuels our lives.

It's not just about climate change, it's about acid rain too. Imagine our skies crying tears of acid, a consequence of pollution from factories and cars. Catalytic converters are like tiny knights in shining armor, battling pollution and cleaning up our air. And by using cleaner fuels and technologies, we can soothe the skies and protect our planet.

Think of the Amazon rainforest as the Earth's lungs, breathing in CO2 and exhaling life-giving oxygen. But deforestation is like a disease, slowly choking our planet. Projects like the Great Green Wall are like a healing balm, restoring degraded lands and planting seeds of hope. Germany's Energiewende is a beacon of change, showing the world that a renewable energy future is within reach.

The Earth is our home, and it's time to give it the care it deserves. By embracing change, planting trees, choosing clean energy, and adopting sustainable practices, we can mend our planet's fever and ensure a vibrant future for generations to come. Let's work together to turn our planet from a sweltering hothouse back into a comfortable, thriving home.

1. Nitrogen Oxides: The Engine's Unwanted Guests (and How to Evict Them)

Imagine your car engine as a bustling city. Fuel and air mix like eager crowds, and then BOOM! – combustion! This fiery fiesta powers your wheels, but it also creates some unwelcome visitors: nitrogen oxides (NOx).

These NOx guys are troublemakers. Think of them as the rowdy partygoers who crash the scene and leave a mess behind. They contribute to smog, which is like a hazy hangover for the city, and acid rain, which is like the morning after's sour stomach for the environment.

But fear not! We have tiny superheroes called catalytic converters living in your car's exhaust system. These microscopic marvels are like bouncers at the engine's nightclub, tossing out the NOx troublemakers before they can cause too much damage.

They do this by performing a neat chemical trick: converting NOx back into harmless nitrogen and oxygen. It's like transforming those rowdy partygoers into well-behaved citizens who quietly slip back into the night.

Of course, even our superhero catalysts have their limits. They get sluggish in the cold (who doesn't?), and they can be poisoned by nasty stuff like lead and sulfur. But scientists are constantly working on new and improved catalysts, so hopefully, one day, we'll have an army of super-efficient bouncers keeping our engines (and our planet) clean!

2. Photosynthesis: Nature's Solar Power Plant

Plants are like tiny green factories, powered by the sun. They have this amazing ability called photosynthesis, which is like their secret recipe for turning sunlight into food.

Imagine a chef in a kitchen, using sunlight to bake a delicious cake. The ingredients? Carbon dioxide from the air and water from the ground. The oven? Chloroplasts, tiny green ovens inside the plant's cells. And the secret ingredient? Chlorophyll, the magical green pigment that captures sunlight like a solar panel.

The chef (the plant) mixes the ingredients (CO2 and water) in the oven (chloroplasts), using the energy from the sunlight (captured by chlorophyll) to bake a delicious cake (glucose). And as a bonus, the chef releases some fresh oxygen into the air, like the sweet aroma of a freshly baked cake.

This "cake," or glucose, is the plant's energy source, its fuel for growth and all its activities. And guess what? When we eat plants (or animals that eat plants), we're basically stealing their cake! Talk about a sweet deal.

But here's the thing: when we cut down forests, it's like destroying those tiny factories. Fewer factories mean less cake and less oxygen, and more CO2 building up in the atmosphere. And that's a recipe for disaster, like a cake that's burnt to a crisp.

3. Photosynthesis: The Equation of Life

Here's the simple recipe for photosynthesis, written in two ways:

Word Recipe: carbon dioxide + water + sunshine → glucose (yum!) + oxygen (ah, breathe it in!)
Chef's Shorthand: 6CO2(g) + 6H2O(l) + light energy → C6H12O6(AQ) + 6O2(g)
This equation is like the secret code of life, the key to understanding how plants create food and sustain life on Earth. It's a reminder that even the most complex processes can be broken down into simple ingredients and reactions. And it's a call to action to protect our planet's green factories, the source of our food, our oxygen, and ultimately, our survival.

In Conclusion:
From the bustling cities of our car engines to the tiny green factories of plants, chemistry and biology are constantly at work, shaping our world in fascinating ways. By understanding these processes, we can appreciate the intricate beauty of nature and work towards a more sustainable future.

Organic chemistry

Formulae, functional groups and terminology

Imagine molecules as tiny, bustling cities. Each atom is a resident with a specific job, and the bonds between them are the roads connecting these residents. Just like a city map helps you navigate, structural formulas and displayed formulas help chemists navigate the intricate world of molecules.

1. Structural Formulas: A Sneak Peek into the City Plan

Think of a structural formula as a quick, concise overview of the city layout. It tells you the key landmarks (atoms) and how they are connected by roads (bonds), but it doesn't show every single house or alleyway.

For example, $CH_2{=}CH_2$ (ethene) is like a small town with two major landmarks (carbon atoms) connected by a superhighway (double bond). Each landmark also has two smaller roads (single bonds) leading to residential areas (hydrogen atoms).

2. Displayed Formulas: A Detailed City Map

Now, let's zoom in! A displayed formula is like a detailed city map, showing every single house (atom) and street (bond). It's like getting a bird's-eye view of the entire city, revealing its intricate network of connections.

Imagine drawing a displayed formula for ethanol (CH_3CH_2OH). You'd start by placing your two carbon landmarks, then carefully draw in all the hydrogen houses and the oxygen landmark with its attached hydrogen. It's like creating a miniature model of the molecule!

Why These Maps Matter

Just like different city layouts create unique urban experiences, different molecular structures lead to diverse properties. These formulas help chemists:

Identify the "neighborhoods" within a molecule: They reveal which atoms are bonded together, like identifying different districts in a city.
Distinguish between "expressways" and "local roads": They differentiate between single, double, and triple bonds, just like highways and smaller roads on a map.
Get a sense of the city's "vibe": While not a 3D model, displayed formulas offer clues about the molecule's overall shape and potential interactions, much like how a city map hints at its overall character.

Beyond the Basics: Exploring Hidden Alleys and Scenic Routes

Just as experienced city explorers venture beyond the main tourist attractions, chemists use advanced representations to delve deeper into molecular structures:

Skeletal Formulas: Like a minimalist map highlighting only major landmarks and routes, skeletal formulas simplify the representation by focusing on the carbon backbone.
3D Representations: These are like interactive 3D city models, allowing chemists to visualize the molecule from all angles and understand its spatial arrangement.
Stereochemistry: Imagine adding extra labels to your city map to indicate one-way streets or specific landmarks. Stereochemistry provides details about the 3D arrangement of atoms around specific points in the molecule.

Case Studies: Unique Cities with Distinct Personalities

Let's explore some fascinating molecular cities:

Butane: Like two cities with the same population but different layouts, butane's isomers (n-butane and isobutane) have distinct properties due to their varying structures.
But-2-ene: Imagine a city with a river running through it, creating two distinct sides. But-2-ene's geometric isomers (cis and trans) arise from the restricted rotation around its double bond, leading to different properties.
Lactic Acid: Like a city with a mirror image twin, lactic acid has optical isomers (L and D) that are non-superimposable mirror images

with unique interactions with light and potentially different biological activities.

In Conclusion

Structural formulas and displayed formulas are like maps that guide chemists through the intricate world of molecules. They provide a visual language to describe molecular architecture, differentiate between isomers, and predict molecular properties. By mastering these representations, we unlock the secrets of the molecular world and gain a deeper appreciation for the diversity and complexity of matter.

Imagine a world where molecules are like LEGO bricks, and organic chemistry is the art of building with them.

Homologous Series: The LEGO Sets of Chemistry

Just like LEGO offers themed sets with different-sized bricks, organic chemistry has "homologous series."

These are families of molecules that share the same basic building block (functional group) but differ in size. Think of alkanes as the basic brick set, alkenes as the one with bendy connectors, alcohols as the set with special water-loving bricks, and carboxylic acids as the ones with both bendy connectors and water-loving parts.

Each time you add a "methylene" brick (-CH2-), you get a slightly bigger molecule with similar properties but slightly different physical traits. It's like adding another car to your LEGO train – it gets longer and heavier, but it's still a train.

Structural Isomers: Rearranging the LEGO Creation

Now, imagine building a LEGO car. You can use the same bricks to create different models, right? That's what structural isomers are! They have the same molecular formula (the same bricks) but different structures (different arrangements).

Chain isomers are like building a long, straight car versus a compact, branched one.

Position isomers are like moving the headlights of your LEGO car to a different spot.
Functional group isomers are like swapping the wheels of your car for wings, turning it into a plane!

Isomerism in Action: The Thalidomide Tale

This isn't just a theoretical game. Isomers can have real-world consequences, as dramatically illustrated by the thalidomide tragedy. This drug, prescribed to pregnant women in the 1950s, existed as two isomers – one helpful, the other harmful. It's like having two LEGO cars that look almost identical but one has a hidden explosive! This tragic incident highlighted the critical importance of understanding isomers in medicine.

The Significance of Molecular LEGO

Understanding homologous series and isomers is like having the instruction manual for building with molecular LEGO. It helps us:

Organize the chaos: Imagine a giant box of LEGO with no organization. That's what organic chemistry would be like without these concepts.
Predict the behavior: Just like you can guess how a LEGO structure will behave based on its design, we can predict the properties of molecules.
Create new wonders: By understanding the rules of molecular LEGO, we can design new materials and medicines with specific properties.
So, the next time you encounter a molecule, remember the LEGO analogy. It might just be the key to unlocking its secrets and building something amazing!

Functional Groups: The Spice of Organic Life

Imagine organic chemistry as a massive, bustling city. Each molecule is like a citizen with its own unique personality. But what gives them that personality? It's their functional groups! These special groups of atoms are like the "hats" molecules wear – they determine how they interact with the world around them.

Think of a hydroxyl group (-OH) as a tiny water balloon, making molecules like ethanol (in your beer!) mix well with water. A carboxyl group (-COOH), on the other hand, is like a little lemon, giving molecules a sour personality, like acetic acid (vinegar!).

Why are functional groups so important?

They're like molecular fingerprints: Just like your fingerprints identify you, functional groups help chemists identify and classify millions of organic compounds.
They're the life of the party: Functional groups determine how molecules react with each other. Want a molecule that explodes? Add a nitro group! Need a molecule that sticks to things? An epoxy group is your friend!
They're the architects of the molecular world: By understanding functional groups, chemists can design new molecules with specific properties, like life-saving drugs, durable plastics, or vibrant dyes.

Homologous Series: Families with a Common Thread

Now, imagine these molecules with different personalities living together in families. These families are called homologous series. They share the same "last name" (functional group) but have different "first names" (number of carbon atoms).

Think of the alkane family. They're the simple, laid-back folks in the organic city. Methane (CH4) is the baby of the family, ethane (C2H6) is the older brother, and so on. They all get along because they share the same "single bond" personality.

What makes a homologous series tick?

Family resemblance: All members have the same functional group, giving them a shared identity and similar chemical behavior.
Growing up together: Each member differs from the next by a -CH2- unit, like adding a building block to a Lego tower.
Personality quirks: As the family grows, their physical properties change gradually. The boiling point increases, making them go from gas to liquid (think of methane gas vs. liquid hexane).

Case Studies: Meet the Families

The Alcohols: This family loves to party and mix with water thanks to their -OH group. They get more "chill" as they grow older (higher boiling points).
The Carboxylic Acids: These are the sourpusses of the organic city, thanks to their -COOH group. They get more and more introverted (less soluble in water) as their carbon chain grows longer.

In the End...

Functional groups and homologous series are the keys to unlocking the secrets of organic chemistry. They help us understand the amazing diversity of molecules and how they interact to create the world around us. So, next time you encounter a molecule, remember to ask: "What's your functional group?" and "Who's in your family?" You might be surprised by what you learn!

Imagine a bustling city:

Saturated Compounds are like the well-organized, predictable streets. They have a clear structure, with every building (carbon atom) connected by single-lane roads (single bonds). Traffic flows smoothly, things are stable, and there's not much room for surprises or changes. Life here is like an alkane – methane, ethane, propane – reliable and steady. Think of them as the sturdy framework of the city.

Unsaturated Compounds are the exciting, dynamic alleyways with hidden shortcuts. They have double or triple bonds, like secret passages (pi bonds) that allow for quick movement and unexpected connections. These alleyways are full of life – alkenes like ethene and propene, and alkynes like ethyne, bring a spark of energy and possibility. They're the vibrant murals and hidden cafes that make the city unique.

Why are these "city streets" important?

Think of unsaturated fats in avocados and olive oil as those lively alleyways. They bring flexibility and vibrancy to our bodies, helping our cells communicate and stay healthy.

Saturated fats in butter and red meat are like the main streets – important but maybe a bit too much of the same. Too many can clog up the city (our arteries), leading to health problems.

Unsaturated compounds are the master builders of our world. They form the long chains (polymers) that make up plastics, from the bottles we use to the clothes we wear.

But they can also be troublemakers. Unsaturated compounds in car exhaust react in the air to form smog, like a hazy cloud hanging over the city.

The story continues:

Scientists are always exploring new ways to use these compounds, like creating new materials or finding better ways to protect our environment. It's like they're constantly redrawing the city map, finding new connections and possibilities in this fascinating world of saturated and unsaturated compounds.

Naming organic compounds

Imagine a world built with Legos. Each Lego brick is a carbon atom, and they love to link up and form chains. Sometimes they hold hands gently (single bonds), and sometimes they cling tightly (double bonds). These chains are the backbone of organic chemistry, the chemistry of life itself!

1. Meet the Unbranched Families

(a) Alkanes: The Laid-Back Bunch

Think of alkanes as the easygoing members of the organic family. They're simple, stable, and content with their single bonds. They're like the long, straight roads of the molecule world.

Methane (CH4): The tiny tot of the alkane family. Just one carbon with four hydrogens clinging on. Like a little baby, it's full of potential energy! 👶
Ethane (C2H6): Two carbons holding hands, each with three hydrogens tagging along. A bit like a happy couple starting their journey. 👫
Propane (C3H8): Three carbons in a row! This is getting exciting. Like a small family, they're ready to fuel your barbecue. 🔥
Butane (C4H10): Four carbons strong! Butane is ready to power your lighter and get the party started. 🎉

(b) Alkenes: The Double Bond Daredevils

Alkenes are a bit more adventurous. They have at least one double bond, which makes them more reactive and ready to shake things up! Think of them as the zip-lines in the molecule world, adding a bit of excitement.

Ethene (C2H4): Two carbons holding on tight with a double bond! This is the building block for lots of plastics, like the ones in your water bottle. ♻️

Propene (C3H6): Three carbons, with a double bond adding a twist. It's used to make polypropylene, a tough plastic found in everything from car parts to yogurt containers.

But-1-ene and But-2-ene (C4H8): Four carbons, but the double bond can be in different places! This is where things get interesting – like siblings with different personalities.

(c) Alcohols: The Life of the Party

Alcohols are like the social butterflies of the organic world. They have an -OH group that loves to interact with other molecules. They're found in everything from drinks to disinfectants.

Methanol (CH3OH): The simplest alcohol, but don't be fooled – it's toxic! Used in things like antifreeze and fuel.

Ethanol (C2H5OH): This is the one you find in your adult beverages. It can make you feel happy (or dizzy!).

Propan-1-ol and Propan-2-ol (C3H7OH): Three carbons and an -OH group, but in different positions! Like isomers, they have the same atoms but different arrangements.

Butan-1-ol and Butan-2-ol (C4H9OH): Four carbons and an -OH, with even more possibilities for isomer fun!

(d) Carboxylic Acids: The Sour Bunch

Carboxylic acids are like the lemons of the organic world. They have a -COOH group that gives them a sour taste. They're found in vinegar, citrus fruits, and even your sweat!

Methanic acid (HCOOH): The simplest carboxylic acid. Ants use it as a defense mechanism – ouch!

Ethanoic acid (CH3COOH): This is what gives vinegar its tang. It's also used to make plastics and glues.

Propanoic acid (C2H5COOH): Found in some cheeses and used as a preservative.

Butanoic acid (C3H7COOH): This one has a strong, unpleasant smell... like rancid butter!

(e) Products of Reactions: The Remix!

This is where the real fun begins! Organic molecules love to react and create new things. It's like a molecular dance party! But to understand the products, we need to know the specific reactions from sections 11.4-11.7. Can you share those details?

2. Unbranched Esters: The Sweet Smell of Success

When alcohols and carboxylic acids get together, they form esters – and release a water molecule in the process. Esters are responsible for many of the pleasant smells in the world, like fruits and flowers.

Imagine it like this: the alcohol and carboxylic acid are holding hands, and when they hug tightly, a water molecule pops out, and they become an ester!

Here are some examples:

Methyl methanolate (HCOOCH3): The simplest ester, with a faint fruity smell.
Ethyl methanolate (HCOOCH2CH3): Smells a bit like rum!
Methyl ethanoate (CH3COOCH3): Has a sweet, fruity aroma like nail polish remover.
Ethyl ethanoate (CH3COOCH2CH3): Smells like pear drops!
And many more! The possibilities are endless, just like the variety of scents in a garden.

Fossil Fuels: The Sun's Ancient Secret, Now Burning in Our Hands

Imagine the Earth millions of years ago. Lush swamps teeming with giant ferns, vast oceans swirling with microscopic life – a world bursting with energy. That energy, captured from the sun, didn't disappear. It was stored, locked away in the bodies of those ancient organisms. Today, we call that stored sunshine "fossil fuels."

A Tale of Three Fuels

Think of fossil fuels as Earth's time capsules, each holding a different kind of energy:

Coal: Imagine layers of ancient forests, pressed and cooked into a dark, earthy rock. We dig up this "rock" (coal) and burn it to power our cities.
Natural Gas: Invisible to the eye, this fuel is mostly methane – a gas that was trapped underground like air in a balloon. We use it to heat our homes and cook our food.
Oil: This liquid gold is the trickiest one. We pump it from deep underground, then refine it into gasoline for our cars, diesel for our trucks, and even the plastic in our toys.

The Methane Mystery

Natural gas is often called a "cleaner" fossil fuel. It burns with less smoke than coal, that's for sure. But here's the catch: it contains methane, a gas that's like a supercharged heat blanket for the planet. Methane traps heat far more effectively than carbon dioxide, making it a major player in climate change.

From Swamp to Spark: The Fossil Fuel Story

How do ancient plants and sea creatures become the fuels we use today? It's a story of time, pressure, and a bit of Earth's magic:

Life and Death: Plants soak up the sun's energy, tiny sea creatures feast on those plants, and everyone eventually dies. Their remains settle on the ground or ocean floor.

Burial and Transformation: Over time, layers of sand and mud bury these remains. The weight and heat turn them into a waxy substance called kerogen.

The Pressure Cooker: Deeper and deeper they go, the heat and pressure building. Kerogen transforms into different types of hydrocarbons – the building blocks of fossil fuels.

Coal: Imagine those ancient forests getting squeezed and baked into a dense, black rock.
Oil and Natural Gas: The tiny sea creatures, under immense pressure, turn into liquid oil or gaseous natural gas.

Unearthing the Past: Extracting Fossil Fuels

Getting these fuels out of the ground is a massive undertaking:

Coal: We dig giant mines, stripping away layers of earth or tunneling deep underground to reach the coal seams.
Oil and Gas: We drill deep wells, sometimes miles into the Earth, and pump the oil and gas to the surface.

The Fossil Fuel Dilemma

Fossil fuels have powered our modern world, but they come at a cost. Burning them releases those ancient stores of carbon back into the atmosphere, changing our climate. We're also facing the reality that these resources are finite – they won't last forever.

A Glimpse into the Future

The energy world is changing. Solar panels, wind turbines, and other clean technologies are capturing the sun's energy directly, without the ancient detour through fossil fuels. The transition won't be easy, but it's a journey we must take to ensure a healthy planet for generations to come.

Imagine a world without petroleum. No cars, no planes, no plastics, no synthetic fabrics. It's hard to fathom, isn't it? Petroleum, also known as crude oil, is the lifeblood of modern society, a complex concoction of hydrocarbons that we've learned to harness for an incredible variety of uses.

Think of crude oil as a kind of "hydrocarbon soup," a mixture of different sized molecules made up of hydrogen and carbon atoms. These molecules range from tiny, light ones like methane (the main component of natural gas) to large, heavy ones like asphalt.

To separate this soup into its useful ingredients, we use a process called fractional distillation. Imagine a giant tower, hotter at the bottom and cooler at the top. When crude oil is heated and vaporized, it rises through this tower, and as it cools, different hydrocarbons condense at different levels. It's like a molecular sorting hat, separating the molecules based on their boiling points.

At the top of the tower, the lightest fractions, like gasoline and kerosene, condense. These are the volatile liquids that fuel our cars and airplanes. Further down, we get diesel fuel for trucks and heating oil for homes. And at the very bottom, the heaviest fractions, like lubricating oil and asphalt, settle out.

But it's not just about separating the components. The real magic happens when we start to manipulate these hydrocarbons, breaking them down and rearranging them to create new materials. This is where the petrochemical industry comes in, transforming petroleum into plastics, synthetic fibers, and countless other products that shape our modern world.

Of course, this process isn't without its challenges. Petroleum is a finite resource, and its extraction and refining can have significant environmental impacts. That's why scientists and engineers are constantly working to improve the efficiency of fractional distillation and develop new technologies to reduce our reliance on fossil fuels.

So, the next time you fill up your car or use a plastic product, take a moment to appreciate the incredible journey of petroleum, from its origins deep beneath the Earth's surface to the countless products that enrich our lives. It's a story of human ingenuity and a testament to our ability to harness the power of nature.

Imagine a bustling city, a melting pot of diverse individuals with unique personalities and skills. That's crude oil for you – a complex mix of hydrocarbon molecules, each with its own "character" defined by its chain length.

Now, picture a skyscraper with floors representing different social strata. This is our fractionating column, where the "social sorting" of hydrocarbons happens. As crude oil is heated and vaporized, it's like these individuals are boarding an elevator to find their place in the city.

At the ground floor, we find the "heavyweights" – long-chain hydrocarbons. These are the established families, deeply rooted and interconnected, requiring a lot of energy (heat) to move. They're the bitumen, the foundation of roads and infrastructure.

Ascending the tower, the chains get shorter, the molecules more agile. These are the young professionals, the entrepreneurs, volatile and ready to vaporize into new ventures. They're the gasoline, fueling our cars and our economy.

Think of viscosity as the "social fluidity" of these groups. The heavyweights at the bottom are like a formal gathering, their long chains intertwined, moving with a viscous grace. As we go higher, the short chains are like a lively party, individuals flowing freely with low viscosity.

This "social structure" of hydrocarbons is crucial for the city's (refinery's) functioning. Just as a city needs the right mix of professionals and workers, a refinery adjusts the "temperature" (operating conditions) to meet the demand for different fractions.

But there's more to this story than just social hierarchy.

Isomerization is like a personality makeover. Even within a "social class" (fraction), molecules can have different structures (isomers). It's like siblings with distinct traits. Refineries use isomerization to refine these traits, enhancing their value, like boosting the "charisma" (octane rating) of gasoline.

Cracking is like breaking down old structures to create new opportunities. It's like urban renewal, where heavy fractions are "redeveloped" into lighter, more valuable ones, increasing the "housing supply" (yield) of gasoline.

And just as a city strives for sustainability, refineries are constantly working to reduce their environmental footprint. Understanding the "social dynamics" of hydrocarbons is key to developing cleaner technologies and ensuring a healthy future for our city (planet).

So, the next time you fill up your car or walk on an asphalt road, remember the intricate "social life" of hydrocarbons and the fascinating process of fractional distillation that makes it all possible.

The Amazing Journey of Crude Oil: From Gooey Mess to Everyday Essentials

Imagine crude oil as a chaotic crowd of diverse characters, each with a unique personality and skillset. To make this unruly bunch useful, we send them through a magical sorting machine – fractional distillation. This process neatly separates the crowd based on their boiling points, kind of like organizing a school dance by age. Let's meet some of the key players and discover their hidden talents:

1. The Energetic Gases: The Life of the Party

This lively bunch, including methane, ethane, propane, and butane, are the smallest and most volatile members of the crude oil family. They're always ready to burst into action, providing:

Warmth and Comfort: Like tiny chefs, propane and butane (aka LPG) whip up delicious meals and keep homes cozy, especially in places where natural gas pipelines haven't reached yet. Think of them as the superheroes of rural kitchens!

Real-world example: In India, the government's "Pradhan Mantri Ujjwala Yojana" scheme is replacing smoky wood fires with clean-

burning LPG, bringing better health and a breath of fresh air to millions of homes.
Industrial Powerhouses: These gases also fuel the fiery furnaces and boilers that power our industries, like the giant kilns that bake cement for our buildings.

Building Blocks for a Plastic World: Ethane, the brainy one of the groups, transforms into ethylene, a key ingredient for creating plastics, synthetic fabrics, and countless other everyday items. It's like the master artist of the chemical world!

Imagine: Your plastic water bottle started its life as ethane, a tiny gas molecule in crude oil. That's molecular magic!

2. Gasoline: The Speedy Fuel

This popular group of hydrocarbons, with 5 to 12 carbon atoms, is the heartthrob of the automotive world.

Road Trip Ready: Gasoline's volatility makes it the perfect fuel for cars and motorcycles. It easily evaporates and mixes with air, creating a powerful explosion that propels our vehicles forward.

Did you know? The "octane rating" of gasoline measures its ability to resist knocking, that annoying engine sound. Higher octane means smoother rides and happier engines.
A Cleaning Companion (with caution!): Gasoline can also act as a solvent to dissolve grease and oil, but it's a bit like a rebellious teenager – powerful but needs to be handled with care due to safety and environmental concerns.

3. Naphtha: The Chemical Chameleon

Naphtha, with 7 to 14 carbon atoms, is the master of disguise in the chemical world.

Transforming into Treasures: Like a magician, naphtha undergoes "steam cracking" to produce ethylene, propylene, and other essential building blocks for plastics, synthetic rubber, and a whole range of chemicals. It's like the ultimate transformer!

Think about it: The plastic in your computer keyboard, the synthetic fibers in your clothes, and even the rubber in your car tires – they all likely started their journey as naphtha.
Octane Booster: Naphtha can also be blended with gasoline to fine-tune its properties, like a skilled bartender creating the perfect cocktail.

4. Kerosene: Taking to the Skies

Kerosene, with 12 to 15 carbon atoms, is the high-flying member of the crude oil family.

Winged Wonder: Kerosene's high flash point and energy density make it the fuel of choice for jet engines, powering our airplanes across continents.

Fun fact: Jet fuel is carefully crafted to withstand extreme temperatures and pressures, ensuring safe and efficient flights.
Light and Warmth: In some parts of the world, kerosene still lights up homes and powers portable stoves, bringing warmth and comfort to remote communities.

5. Diesel: The Powerful Workhorse

Diesel, with 15 to 25 carbon atoms, is the strong and reliable member of the crude oil family.

Heavy Duty Hero: Diesel engines, commonly found in trucks, buses, and heavy machinery, rely on diesel's ability to ignite under pressure. It's the engine behind our global transportation and construction industries.

Did you know? The "cetane number" of diesel measures its ignition quality. Higher cetane means smoother engine operation and cleaner emissions.
Keeping Things Warm: Diesel also keeps us warm by fueling industrial boilers and furnaces, like the ones that produce steel and glass.

6. Fuel Oil: The Industrial Giant

Fuel oil, with its heavy hydrocarbons, is the gentle giant of the crude oil family.

Powering the World: Fuel oil powers massive ships that transport goods across oceans and generates electricity in power plants, keeping our world connected and illuminated.

Imagine: The clothes you wear, the food you eat, and the electronics you use – they've likely traveled across the sea on a ship powered by fuel oil.
Industrial Heat: Fuel oil also provides the intense heat needed for industrial processes like steelmaking and glass production.

7. Lubricating Oil: The Smooth Operator

Lubricating oil, with its high viscosity, is the peacekeeper of the mechanical world.

Reducing Friction: Like a microscopic layer of marbles, lubricating oils reduce friction between moving parts in machines and engines, preventing wear and tear and ensuring smooth operation.

Think about it: Your car engine would quickly overheat and seize without the soothing touch of lubricating oil.
Wax On, Wax Off: Paraffin wax, extracted from lubricating oil, is used in candles, packaging, and even surfboards!

8. Bitumen: The Road Builder

Bitumen, the heaviest member of the crude oil family, is the foundation of our modern infrastructure.

Paving the Way: Bitumen is the sticky black magic that binds aggregates together to create durable and waterproof roads. It's the backbone of our transportation networks.

Did you know? Asphalt, a mixture of bitumen and aggregates, is the most common material used for road paving. It's flexible, durable, and can withstand heavy traffic.
Keeping Things Dry: Bitumen also protects our roofs and foundations from water damage, ensuring our buildings stay safe and dry.

The End of the Journey (and a New Beginning)

Crude oil's journey from a messy mixture to a diverse range of valuable products is a testament to human ingenuity. But it's important to remember that crude oil is a finite resource, and its extraction and use have environmental consequences. As we continue to rely on these petroleum fractions, it's crucial to explore sustainable alternatives and promote energy efficiency to ensure a cleaner and brighter future for generations to come.

Alkanes

Imagine the bustling metropolis of Organic Chemistry. In this sprawling city, with its complex structures and endless reactions, there's a quiet neighborhood where it all begins: Alkane Alley.

These are the founding families, the simple folk of the organic world. No flashy double or triple bonds for these guys – just good, honest single bonds between their carbon and hydrogen residents. Think of them as the strong, silent types, content with their straightforward lifestyle. This "single life" makes them pretty laid-back; they're not the type to get involved in a lot of drama (reactions). But, like everyone else, they have their moments – lighting up the barbecue (combustion) or getting into a little chlorine-induced spat (halogenation).

Let's get to know these Alkane Alley residents a bit better:

Bonding: Picture each carbon atom as a tiny house with four windows (orbitals). They love symmetry in Alkane Alley, so these windows are evenly spaced, forming a perfect tetrahedron. Each window connects to another house (carbon) or a simple hydrogen home, creating a strong and stable neighborhood. No gossip or electron-hoarding here – everyone shares nicely!

Structure: Alkane Alley boasts a diverse range of architecture. Some houses line up in neat rows (straight-chain alkanes), while others branch out with quirky extensions (branched-chain alkanes). And then there are the cul-de-sacs, where the houses form a cozy circle (cyclic alkanes). It's this variety that gives Alkane Alley its unique charm.

Properties: Life in Alkane Alley is pretty chill. Melting points and boiling points are low, so things stay relaxed even when the temperature rises. They're not big fans of water (who needs those polar personalities anyway?), preferring to hang out with their nonpolar pals like benzene and ether. And they're always rising to the top (less dense than water, you see).

Reactions: As we mentioned, alkanes are generally peaceful folks. But even the quietest neighborhood has its occasional excitement:

Combustion: When things get heated, alkanes know how to throw a party! They're excellent fuels, releasing a burst of energy (and a bit of CO2 and water) when they combust. Think of it as their way of letting loose and having a good time.

Halogenation: Sometimes, those mischievous chlorine atoms come sneaking around, looking to stir up trouble. They'll try to replace a hydrogen resident, causing a bit of a commotion. But hey, even the most stable neighborhoods have their occasional squabbles.

Alkane Alley in Action:

Natural Gas: This is the energy hub of Alkane Alley, mainly powered by methane and his ethane, propane, and butane buddies. They keep things running smoothly (and cleanly!) in many homes and industries.

Petroleum: This is the "big city" just beyond Alkane Alley, a diverse mix of hydrocarbons. Alkanes play a major role here, fueling our cars and planes.

Polyethylene: Okay, so this one technically involves a relative from Alkene Avenue (ethene), but it's worth a mention. Ethene's ability to link up and form long chains (polyethylene) is what gives us those handy plastic bottles and bags.

Chloroform: Remember those chlorine troublemakers? Well, they're responsible for creating chloroform, a once-popular anesthetic. It's a bit too potent for regular use now, but it still finds work as a solvent and chemical building block.

So, there you have it – a glimpse into the world of alkanes. They may be simple, but they're essential to the fascinating world of organic chemistry. Think of them as the reliable foundation upon which the rest of the city is built.

Imagine a dance floor...

...but instead of people, it's filled with tiny alkane molecules, holding hands and happily bouncing around. These are pretty chill molecules, content in their simple existence. Suddenly, the lights flash on, pulsing with UV energy! This is no ordinary disco ball – it's showering the dance floor with energetic photons. Enter the chlorine molecules, sleek and powerful, like a pair of dancers ready to break it down. The UV light zaps them with energy, splitting them apart into solo dancers – chlorine radicals. These radicals are wild and ready to mingle, each with an unpaired electron making them desperate to find a partner.

The music starts pumping...

...and the substitution reaction begins! A chlorine radical zooms towards a methane molecule, grabs a hydrogen atom's hand, and swings it away into a passionate dance move (forming HCl). This leaves the methane molecule a bit unsteady, now a methyl radical with a lonely unpaired electron. But fear not! Another chlorine molecule glides in, ready to sweep the methyl radical off its feet. They bond, forming chloromethane, and a chlorine radical is spun off to continue the frenzy. This electrifying exchange keeps happening – radicals stealing partners, new molecules forming, and the dance floor buzzing with energy.

But like all good parties, this one has to end...

Eventually, two radicals bump into each other and decide they've had enough excitement. They form a stable molecule, the music fades, and the dance floor calms down. The result? A mix of new molecules, some with a single chlorine atom replacing a hydrogen (Mon substitution), and others with multiple chlorine partners. It's like a molecular mixer, with new relationships and possibilities formed.

Why should we care about this chemical dance-off?

Well, these chlorinated alkanes are more than just party animals. They're versatile molecules with real-world jobs:

Chloroform: Once a dramatic star in the operating room (putting patients to sleep for surgery), it now prefers a quieter life as a solvent and cleaning agent.
CFCs: These cool cats were the life of the refrigeration party, but their ozone-depleting moves got them banned. They've been replaced by more eco-friendly dancers.

The takeaway?

Substitution reactions, like a wild dance party, show us how even seemingly simple molecules can transform when given the right conditions. By understanding these chemical moves, we can create new materials and technologies, but we also need to be mindful of the consequences and dance responsibly.

Alkenes

Alkenes: The Spice of Hydrocarbon Life

Imagine the world of hydrocarbons as a grand symphony. The alkanes, with their steady single bonds, are the reliable bass line – always there, always predictable. But then, the alkenes arrive, like a burst of trumpets, adding a vibrant and exciting melody. These unsaturated hydrocarbons, with their characteristic double bonds, bring a whole new level of complexity and possibility to the chemical world.

The Double Bond: A Dance of Electrons

At the heart of every alkene is the captivating double bond. Picture it as a pair of dancers, twirling and swirling in a graceful partnership. One dancer, the sigma (σ) bond, represents a strong, head-on embrace, while the other, the pi (π) bond, is a more delicate, side-by-side connection. This dynamic duo creates a rigid structure, preventing the free rotation that characterizes alkanes.

Unsaturated and Ready to React

Alkenes are like those eager party guests who always seem to be in the middle of the action. Their "unsaturated" nature means they have fewer hydrogen atoms than their alkane counterparts, leaving them with a craving for more. This makes them highly reactive, readily participating in a variety of chemical reactions.

Geometric Isomers: Mirror Images with a Twist

The double bond also introduces the intriguing concept of geometric isomerism. Imagine two molecules that are mirror images of each other, yet somehow different. These are cis-trans isomers, and they arise because the double bond locks the atoms in place, preventing them from rotating freely. When similar groups are on the same side of the double bond, we have the cis isomer, like two dancers facing each other. When they're on opposite sides, it's the trans isomer, like dancers turning away from each other.

Cracking the Code: From Big to Small

In the bustling refinery, alkenes are born from a process called cracking. It's like taking a giant Lego creation and breaking it down into smaller, more versatile pieces. Thermal cracking uses intense heat to shatter the bonds of large alkane molecules, while catalytic cracking employs clever catalysts to achieve the same result with less energy.

Why Crack? The Benefits of Breaking Down

Cracking isn't just about destruction; it's about transformation. By breaking down large alkanes, we can:

Match supply and demand: Crude oil often contains an excess of long-chain alkanes. Cracking helps us convert these into the shorter-chain alkanes and alkenes that are in high demand.
Create building blocks: Alkenes are the essential ingredients for countless products, from plastics and synthetic fibers to pharmaceuticals and detergents.
Boost fuel quality: Cracking can improve gasoline by increasing the proportion of branched-chain alkanes, which burn more smoothly in engines.
Produce hydrogen: This valuable byproduct is used in everything from fertilizer production to clean energy technologies.

Fluid Catalytic Cracking: A Refinery Star

One of the most impressive cracking processes is fluid catalytic cracking (FCC). It's like a choreographed dance of catalyst and hydrocarbons, swirling together in a reactor to produce a stream of valuable products. FCC is a workhorse of the refinery, helping to meet the world's insatiable demand for gasoline and other fuels.

Alkenes: The Future is Unsaturated

As we move towards a future where sustainability and resourcefulness are paramount, alkenes will continue to play a vital role. Their versatility and reactivity make them indispensable in the creation of new materials and technologies. So, let's celebrate these dynamic hydrocarbons, the ones that bring a touch of spice and excitement to the world of chemistry.

Imagine a bustling city. Its roads, representing saturated hydrocarbons, are filled with cars, each carrying the maximum number of passengers (hydrogens). Traffic flows smoothly, everyone content in their designated spots.

Now picture a coastal highway. This is our unsaturated hydrocarbon, with cars (carbons) zooming along, some with empty seats (representing the potential for more hydrogens). Suddenly, a lively group of surfers (bromine) arrives, eager to join the ride. They hop into the available seats in the cars with empty spots, changing the dynamic of the journey. The once-empty highway now teems with a new energy, the surfers adding their vibrant presence to the mix.

This is the bromine water test in action. The surfers (bromine) readily join the cars with empty seats (unsaturated hydrocarbons), changing the color of the solution as they disappear from the water. But the cars already full of passengers (saturated hydrocarbons) continue on their way, unaffected by the surfers' presence.

Think of addition reactions as building a LEGO tower. Each brick (molecule) snaps onto another, creating a single, unified structure (product). Unsaturated hydrocarbons, with their open connection points (double or triple bonds), are like LEGO bricks with extra knobs, ready to connect and build.

Alkenes, with their double bonds, are the daredevils of the hydrocarbon world. They're always ready for an adventure, eager to react and form new connections.

Picture them bungee jumping (addition of bromine) into a pool of bromine water, the bromine molecules latching onto them as they plunge.
Or imagine them skydiving (hydrogenation) into a cloud of hydrogen, the hydrogen molecules clinging on for the ride, transforming the alkene into a more grounded alkane.
Finally, envision them riding a rollercoaster (hydration) through a misty tunnel of steam, the water molecules joining them for the exhilarating journey, creating an alcohol as they emerge.
The polymerization of ethene is like a conga line at a lively party. Each person (ethene molecule) joins hands (breaks its double bond) with another, forming a long, snaking chain (polyethylene). The music pumps, the line grows, and the party gets even more exciting!

This is the magic of chemistry, where molecules dance and transform, creating the world around us.

Imagine a tiny, bustling city populated by alcohol molecules. These aren't your stuffy, old-fashioned alcohols, mind you. These are vibrant, energetic characters with a fiery spirit. They love to dance, and their favorite dance partner is oxygen.

When the music starts (that's the heat!), the alcohols and oxygen molecules twirl and tango, getting closer and closer. Suddenly, they embrace in a passionate burst of energy – that's combustion! This fiery dance releases a joyous chorus of light and heat, leaving behind a wisp of steam (water) and a satisfied sigh (carbon dioxide).

The Longer the Chain, the Wilder the Dance

Now, not all alcohols dance the same way. Some, like methanol (the smallest of the bunch), are quick and nimble, igniting with a bright, clean flame. Think of them as the nimble ballerinas of the alcohol world.

But as the alcohol molecules get bigger (ethanol, propanol, and so on), their dance becomes more complex and dramatic. These are the fiery flamenco dancers, their long carbon chains swirling and stamping with passion. Sometimes, if they don't have enough oxygen partners, their dance gets a bit smoky and chaotic – that's incomplete combustion.

Ethanol: The Life of the Party

Ethanol, the life of the party, is a true multi-talent. Not only is it a fantastic dancer (burning clean and efficiently), but it's also a master of disguise!

The Solvent Superhero: Ethanol can dissolve almost anything, from the perfumes that make us smell divine to the medicines that keep us healthy. It's like the ultimate social butterfly, blending in seamlessly with every crowd.

The Eco-Warrior: Tired of fossil fuels polluting the planet? Ethanol steps in as the eco-friendly fuel, produced from plants like corn and sugarcane. It's the superhero we need to fight climate change, one clean-burning engine at a time.

The Future is Burning Bright

The story of alcohol combustion is still being written. Scientists are working on exciting new chapters, like finding ways to make ethanol from things like wood chips and algae. Imagine a world where fuel grows on trees!

So, the next time you see a flame flicker, remember the passionate dance of the alcohol molecules. It's a story of energy, transformation, and the power of chemistry to light up our world.

Ethanol Production: A Tale of Two Methods

Imagine ethanol as a versatile hero in the world of chemistry, playing roles as diverse as a cleaning agent, a germ-fighter, a fuel booster, and the life of the party in your favorite adult beverage. But where does this hero come from? There are two main origin stories: the ancient art of fermentation and the modern marvel of hydration.

Fermentation: The Ancient Alchemist

Picture this: a bubbling cauldron, not filled with a witch's brew, but with a sugary concoction of fruits or grains. This is the heart of fermentation, a process older than written history, where tiny organisms called yeast work their magic.

Yeast: The Microscopic Magician: Yeast cells are like tiny alchemists, possessing enzymes that can transform sugar into ethanol and carbon dioxide. It's like they have a secret recipe passed down through generations!
No Air Allowed: This magic happens only in the absence of oxygen. If oxygen were present, it would be like throwing water on the alchemist's fire, resulting in nothing but carbon dioxide and water – no ethanol for our hero's journey!
Just the Right Temperature: The cauldron needs to be kept at a cozy temperature, not too hot, not too cold (around 25-35°C). This keeps the yeast happy and productive, ensuring a steady supply of ethanol.

The Fermentation Spell book:

Glycolysis: The Sugar-Splitting Ritual: In this first step, glucose, the sugar molecule, is broken down into smaller pieces called pyruvate. It's like chopping up ingredients before cooking a grand feast.
Pyruvate's Transformation: Now, the pyruvate undergoes a magical transformation. It loses a bit of itself as carbon dioxide and then, with the help of a special ingredient called NADH, it's reborn as ethanol!

From Cauldron to Industry:

Gathering the Ingredients: Just like any good recipe, fermentation starts with the right ingredients. Sugarcane juice, corn starch, or molasses are the usual suspects, providing the sugary base for the yeast to work on.
The Brewing Begins: The sugar solution is mixed with yeast and left to ferment in large vats, carefully monitored for temperature and other conditions. It's like a giant brewery, but instead of beer, we're brewing ethanol!
Distillation: The Refining Fire: The fermented broth, now containing ethanol, is put through a fiery trial called distillation. This separates the ethanol from the water and other impurities, increasing its concentration.
Dehydration: The Final Touch: To get the purest form of ethanol, a final spell called dehydration is cast. This removes any remaining traces of water, leaving us with nearly 100% pure ethanol.

The Pros and Cons of Fermentation:

Pros:

Nature's Bounty: Fermentation relies on renewable resources like sugarcane and corn, making it a sustainable choice.
Energy-Efficient: Compared to the hydration method, fermentation is like a gentle simmer rather than a roaring fire, requiring less energy.
Time-Tested: This process has been used for centuries, so we know it works!

Cons:

Batch by Batch: Fermentation is like baking a cake – you have to do it in batches, which can be slower than a continuous process.
Weak Brew: The initial ethanol concentration is low, requiring extra effort to concentrate it.
Food vs. Fuel: Using food crops for ethanol production can raise ethical concerns, as it diverts resources from feeding people.

Hydration: The Modern Alchemist

Now, let's step into a modern laboratory, where instead of bubbling cauldrons, we have gleaming reactors and pipelines. This is the realm of hydration, a chemical process that transforms ethene, a gas derived from petroleum or natural gas, into ethanol.

Ethene: The Gaseous Starting Point: Ethene is like the raw material, the building block for our ethanol hero.
Steam and Catalyst: The Magical Reagents: Ethene is mixed with steam and a special ingredient called a catalyst, usually phosphoric acid. This catalyst acts like a magical guide, helping the reaction along.
Heat and Pressure: The Forging Fire: The reaction takes place under high heat (300°C) and pressure (60 atmospheres), like forging a sword in a fiery furnace.

The Hydration Spell book:

Protonation: The Awakening: The catalyst awakens the ethene molecule, making it ready to react.
Water's Embrace: Water molecules, like eager dancers, join the ethene, forming a new bond.
Deprotonation: The Completion: The catalyst steps back, leaving behind a newly formed ethanol molecule.

From Lab to Industry:

Ethene Extraction: Ethene is obtained from natural gas or petroleum, like mining precious ores from the earth.

The Reaction Chamber: Ethene and steam are mixed and passed over the catalyst in a high-temperature, high-pressure reactor. It's like a magical chamber where the transformation takes place.
Separation and Purification: The resulting mixture is cooled and separated, and the ethanol is further purified to remove any remaining impurities.

The Pros and Cons of Hydration:

Pros:

High Purity: Hydration produces highly pure ethanol, often exceeding 99%.
Continuous Flow: This process runs continuously, like a river, allowing for efficient production.
Non-Food Feedstock: Ethene comes from fossil fuels, so it doesn't compete with food production.

Cons:

Fossil Fuel Reliance: Hydration depends on fossil fuels, which are non-renewable and contribute to climate change.
Energy-Intensive: The high temperatures and pressures require a lot of energy, like keeping a furnace burning constantly.
Catalyst Care: The catalyst can be sensitive and needs to be replaced or regenerated periodically.

The Choice: Fermentation vs. Hydration

So, which method is better? It's like choosing between two heroes with different strengths and weaknesses. Fermentation is the eco-friendly option, relying on renewable resources and requiring less energy. Hydration, on the other hand, produces higher purity ethanol and runs more efficiently. The best choice depends on factors like feedstock availability, desired purity, energy costs, and environmental concerns.

As the world seeks sustainable solutions, fermentation is gaining popularity. But hydration remains a vital process for producing high-purity ethanol for various applications. Ultimately, both methods play a crucial role in providing us with this versatile chemical hero.

Carboxylic Acids: The Sourpusses of the Chemistry World

Imagine a molecule with a split personality. On one side, it's got a carbonyl group (C=O), all prim and proper, like a tightly wound chaperone. On the other, there's a hydroxyl group (-OH), a bit of a wild child, ready to party. Together, they form the carboxyl group (-COOH), the defining feature of carboxylic acids. These molecules are the sourpusses of the chemistry world, lending their tang to everything from vinegar to spoiled milk.

But don't let their sour disposition fool you, carboxylic acids are quite the social butterflies! They love to react with other elements and compounds, leading to some fascinating transformations.

1. Carboxylic Acids and Their Entourage

(a) Metals: A Bubbly Encounter

Just like a grumpy old miser who secretly enjoys a good party, carboxylic acids react with reactive metals like sodium, potassium, and magnesium. Imagine these metals as the life of the party, bursting in with their energetic electrons. The result? A lively exchange where the metal loses its electrons to the carboxylic acid, forming a salt and releasing hydrogen gas – the bubbles that make this chemical soirée fizz!

Think of it like this:

Ethanoic acid (vinegar) + Sodium (a feisty metal) → Sodium ethanoate (a salty character) + Hydrogen gas (the life of the party)

(b) Bases: A Neutralizing Embrace

When carboxylic acids meet bases like sodium hydroxide, it's like a clash of titans. The acidic nature of the carboxylic acid is neutralized by the base, creating a salt and water. It's a classic case of opposites attract, resulting in a peaceful resolution (and a refreshing drink of water).

Picture this:

Ethanoic acid (the sourpuss) + Sodium hydroxide (the peacemaker) → Sodium ethanoate (a neutral bystander) + Water (a refreshing outcome)

(c) Carbonates: A Gassy Get-Together

Carboxylic acids also get along famously with carbonates like sodium carbonate. This reaction is like a baking soda volcano, producing a salt, water, and carbon dioxide gas. The carbon dioxide bubbles add a bit of excitement to this otherwise calm gathering.

Visualize it:

Ethanoic acid (the vinegar) + Sodium carbonate (baking soda) → Sodium ethanoate (a salty surprise) + Water (a necessary ingredient) + Carbon dioxide (the bubbly entertainment)

2. The Birth of Vinegar: A Tale of Oxidation

Ethanoic acid, the main component of vinegar, has a dramatic origin story. It starts with ethanol (the alcohol in your favorite adult beverage) and involves a process called oxidation.

(a) Potassium Manganate (VII): The Purple Alchemist

Imagine potassium manganate (VII) as a powerful wizard with a purple robe. This oxidizing agent, in the presence of an acid, transforms ethanol into ethanoic acid. As the reaction unfolds, the wizard's purple robe fades, signifying the magic at work.

(b) Bacteria: The Tiny Brewers

In the world of vinegar production, bacteria from the Acetobacter genus are the real heroes. These microscopic brewers use oxygen from the air to oxidize ethanol, creating ethanoic acid. It's a slow and steady process, like brewing the perfect kombucha, but the result is a tangy delight.

3. A Love Story: Carboxylic Acids and Alcohols

Carboxylic acids have a special affinity for alcohols. In the presence of an acid catalyst (a matchmaker, if you will), they form esters, compounds known for their delightful fragrances. This reaction, called esterification, is like a love story, where two molecules come together to create something beautiful and aromatic.

Imagine this:

Ethanoic acid (the vinegar) + Ethanol (the alcohol) → Ethyl ethanoate (a fruity fragrance) + Water (a byproduct of their love)
Biodiesel: A Modern-Day Alchemy

This love story has a practical application in the production of biodiesel. Vegetable oils and animal fats, rich in triglycerides (molecules with three ester bonds), undergo a transesterification reaction with methanol. This process, aided by a base catalyst, breaks down the triglycerides and creates fatty acid methyl esters (FAMEs), the building blocks of biodiesel. It's like transforming kitchen waste into liquid gold!

The End (or is it?)

The world of carboxylic acids is full of surprises. Their reactions are essential to countless processes, from the production of everyday products to the development of new technologies. So next time you encounter the sour taste of vinegar or the sweet scent of a flower, remember the fascinating chemistry of carboxylic acids at play.

Polymers: The Unsung Heroes of Our Material World

Imagine a world without plastics, without fabrics that stretch, without the very DNA that makes you, you. Sounds impossible, right? That's because polymers, those magnificent macromolecules, are the silent forces shaping our reality.

Think of them as nature's (and science's) Lego bricks. Tiny building blocks called monomers link together in endless chains, creating everything from the softest silk to the toughest Kevlar. The name itself is a giveaway – "poly" meaning "many" meaning "parts" – like a microscopic conga line of molecules!

Nature's Polymers: Life's Original Masterpieces

Long before humans started tinkering in labs, nature was a master polymer chemist. Our very existence relies on them:

Proteins: The workhorses of our cells, built from amino acid chains. They're like tiny machines, transporting oxygen, fighting invaders, and even making your hair shine.
Cellulose: The sturdy stuff of plants, made from sugar molecules. It gives trees their strength and provides us with paper, cotton, and even the delicious crunch in celery.
DNA: The blueprint of life itself, a twisted ladder of genetic code. It's a polymer so intricate and elegant, it makes even the most complex human inventions look like child's play.

Synthetic Polymers: Humans Join the Party

Not content with nature's handiwork, humans decided to create some polymers of their own. And boy, did we get creative!

Polyethylene: The humble hero of plastic bags and bottles. It's everywhere, for better or worse, a testament to our love-hate relationship with this versatile material.

Nylon: From stockings that revolutionized fashion to parachutes that saved lives, nylon's strength and flexibility have made it a true icon of the synthetic world.
Teflon: The "miracle" coating that keeps your eggs from sticking and your spaceship heat-resistant. Who knew a polymer could be so slippery?

The Polymer Family Tree: Addition vs. Condensation

Polymers, like families, have their own unique dynamics. There are two main branches:

Addition Polymers: Imagine a bunch of monomers holding hands, each contributing a bond to form a long chain. No atoms are lost in this happy union. Think polyethylene and polypropylene – simple, straightforward, and incredibly useful.
Condensation Polymers: These are a bit more dramatic. When monomers join, they lose a tiny bit of themselves (like a water molecule), creating a by-product. But the result is a strong, tightly-knit family, like polyester, nylon, and polycarbonate.

Cracking the Code: Deducing Polymer Structures

Want to play polymer detective? It's easier than you think! If you know the monomer, you can predict the polymer structure. Break the double bonds, link them up, and voila! You've got your repeating unit. It's like a molecular puzzle, and once you get the hang of it, you'll be deciphering polymers like a pro.

Case Studies: Polymer Superstars

Let's meet some polymer celebrities:

Polyethylene: The king of plastics, used in everything from shopping bags to artificial hips. But its reign is not without controversy, as we grapple with its environmental impact.
Nylon: A fashion icon and engineering marvel, nylon's strength and versatility continue to inspire. But its production can be energy-intensive, reminding us that even the most wondrous materials have a footprint.

Kevlar: The superhero of polymers, saving lives with its incredible strength and heat resistance. From bulletproof vests to spacecraft, Kevlar is pushing the boundaries of what's possible.

The Future of Polymers: Innovation and Sustainability

As we continue to explore the world of polymers, we face both exciting possibilities and pressing challenges. How can we create polymers that are both high-performing and environmentally friendly? Can we mimic nature's elegance to design materials that are truly sustainable? The answers lie in the hands of the next generation of polymer scientists, ready to shape a future where these remarkable macromolecules continue to improve our lives.

Condensation Polymerization: A Molecular Dance

Imagine tiny molecular dancers, twirling and pairing up on a grand ballroom floor. As they join hands, they release a tiny puff of steam – a water molecule – a whisper of their energetic connection. This, my friend, is the essence of condensation polymerization.

Unlike their rambunctious cousins in addition polymerization, who simply chain themselves together, these dancers are more refined. They shed a bit of themselves to forge a stronger bond, creating long, elegant chains called polymers.

Polyamides and Polyesters: A Tale of Two Polymers

Let's meet the stars of our show: the polyamides and polyesters.

Polyamides: Think of nylon, the strong and versatile fabric used in everything from stockings to parachutes. It's born from the union of a dicarboxylic acid and a diamine. Picture them as dancers with two arms each (functional groups), reaching out to form a tight embrace (amide linkage).

Polyesters: Now, envision the smooth, shimmering allure of PET, the plastic used in soda bottles and clothing fibers. It's the lovechild of a dicarboxylic acid and a diol, their graceful movements culminating in an ester linkage.

Deducing the Dance Moves:

Like any good dance, there's a pattern to follow. By understanding the steps, we can predict the final choreography (polymer structure) or even reverse engineer it to identify the original dancers (monomers).

Addition vs. Condensation Polymerization: A Dance-Off

While both types of polymerizations create polymers, their styles are distinct:

Addition Polymerization: A high-energy, free-for-all where monomers latch onto each other without losing any parts. It's like a conga line that keeps growing.

Condensation Polymerization: A more intimate affair, where monomers sacrifice a bit of themselves to create a stronger bond. It's like a waltz, where each pair becomes one.

Plastics: The Polymer Playground

Plastics are the tangible result of this molecular dance. They're the toys, the bottles, the fabrics that fill our world. Each plastic's unique properties – its flexibility, strength, and melting point – are determined by the type of polymer it's made from.

Case Studies: Real-World Rhythms

PET Recycling: Like a never-ending dance, PET bottles can be broken down and reborn as new products through depolymerization.

Bio-based Polymers: Nature joins the dance floor with bio-based polymers made from renewable resources like corn starch.

Kevlar: This super-strong polyamide, used in bulletproof vests, owes its resilience to the tight embrace of its polymer chains.

Conclusion: The Dance Goes On

Condensation polymerization is a symphony of molecular movement, creating a diverse array of polymers that shape our world. As we continue to explore this fascinating dance, we'll unlock new possibilities for materials that are stronger, more sustainable, and more attuned to our needs.

The Plight of the Plastic: A Story of Durability and Despair

Imagine a material so resilient it could outlive civilizations, so adaptable it forms everything from life-saving medical devices to disposable coffee cups. That's plastic – a marvel of modern engineering, yet a growing menace to our planet.

1. The Double-Edged Sword of Durability

Plastic's strength is also its downfall. Like a stubborn guest who overstays their welcome, plastic lingers in our environment for centuries, refusing to break down. This "gift" of longevity turns into a curse:

Landfill Bullies: Imagine landfills as bustling cities, and plastics as those residents who refuse to move out. They gobble up precious space, leaving little room for newcomers. Worse still, they leak harmful chemicals like grumpy neighbors spreading gossip, poisoning the surrounding soil and water.

Ocean Drifters: Lightweight and carefree, plastics embark on unplanned adventures, riding ocean currents and forming massive "garbage patches" – swirling vortexes of plastic debris. These plastic islands are death traps for marine life, who mistake them for food or become entangled in their plastic embrace.

Recycling Roadblocks: Plastics are a diverse bunch, each with its own unique personality. This makes sorting and recycling them as complex as organizing a party for guests who speak different languages. Many recycling facilities are ill-equipped to handle this plastic melting pot, leaving a significant portion destined for the landfill.

Case Study: The Epic Journey of a Plastic Bottle

Picture a plastic water bottle, fresh off the production line, eager to quench someone's thirst. It fulfills its duty, but its journey doesn't end there. Tossed aside, it embarks on an odyssey: tumbling through storm drains, drifting down rivers, and finally reaching the ocean. Years turn into decades, then centuries. The bottle, once pristine, now fragments into microplastics – tiny plastic specks that infiltrate the food chain, threatening marine life and ultimately, ourselves.

2. Plastic's Environmental Toll: A Tale of Three Tragedies

Plastic's environmental impact is a three-act play; each act more devastating than the last:

Act 1: Landfill Overflow: Landfills are bursting at the seams, thanks to plastic's refusal to decompose. This forces us to seek new dumping grounds, often encroaching upon pristine natural habitats.
Act 2: Ocean Invasion: The ocean, once a symbol of vastness and purity, is now littered with plastic debris. Marine animals become entangled in plastic nets, ingest microplastics, and suffer the consequences.
Act 3: Toxic Fumes: Desperate to get rid of plastic, we sometimes resort to burning it. But this "solution" releases harmful pollutants, choking the air and contributing to climate change.

Case Study: The Plastic Bag and the Sea Turtle

A sea turtle glides through the ocean, its graceful movements a testament to millions of years of evolution. Suddenly, it spots a jellyfish – a favorite treat. But this "jellyfish" is actually a plastic bag, carelessly discarded by a human. The turtle swallows it, and the bag lodges in its gut, slowly leading to a painful death. This tragic tale, repeated countless times, highlights the devastating impact of plastic pollution on marine life.

3. The Molecular Makeup of the Enemy:

To combat plastic pollution, we need to understand its structure. Let's delve into the molecular world of two common plastics:

Nylon: The Chain of Strength: Imagine a chain of interconnected links, each link representing an amide bond. This is the backbone of nylon, a polymer known for its strength and durability. It's this very strength that makes nylon so persistent in the environment.
PET: The Ring of Resilience: PET, the plastic commonly used in bottles, has a different structure. It features ester linkages and a rigid aromatic ring, contributing to its resilience and resistance to degradation.

The Final Act: A Call to Action

Understanding the properties and structure of plastics is the first step towards finding sustainable solutions. We need to reduce our reliance on single-use plastics, improve waste management systems, and develop innovative recycling technologies. The future of our planet depends on it.

Plastic Fantastic: The Epic Journey of a PET Bottle

Imagine a bubbly bottle of soda, chilled and ready to quench your thirst on a hot summer day. That bottle, my friend, is likely made of polyethylene terephthalate, or PET for short. But this isn't just any ordinary plastic. PET is a superhero in disguise, with a secret power: recyclability!

Think of it like this: PET is a chain of tiny LEGO bricks, all linked together. When it's time to recycle, we have amazing machines that act like tiny LEGO dismantlers, carefully separating those bricks (called monomers). These "loose bricks" are then ready to be rebuilt into something brand new – maybe another bottle, a cozy fleece jacket, or even a super-strong rope!

This incredible process, known as depolymerization, is like a magic trick. We can use special chemicals or even tiny helpers called enzymes to break down the PET. It's like having a microscopic recycling crew working tirelessly to give those LEGO bricks a new life!

Why is this so important? Well, imagine a world overflowing with plastic waste, piling up in landfills and polluting our oceans. PET recycling helps us fight this villain! By reusing those "LEGO bricks," we save energy, protect our planet, and reduce our reliance on fossil fuels. It's a win-win-win!

Of course, even superheroes face challenges. Sometimes, those "LEGO bricks" get mixed with other materials, making it tricky to separate them. But fear not! Scientists are constantly working on new technologies to make PET recycling even more efficient and sustainable.

So, next time you enjoy a refreshing drink from a PET bottle, remember it's incredible journey. It's a story of transformation, resilience, and the power of recycling to create a better world.

Proteins: The Microscopic Machines That Make Life Possible

Imagine tiny, intricate machines, buzzing with activity inside every cell of your body. These microscopic marvels are called proteins, and they're the unsung heroes of life itself.

Built from tiny building blocks called amino acids, proteins are like complex LEGO creations, each with a unique shape and purpose. Some are like speedy delivery trucks, transporting molecules throughout the cell. Others are like skilled construction workers, building and repairing tissues. And some are like master chefs, catalyzing chemical reactions that keep our bodies running smoothly.

These amino acid "LEGO bricks" come in 20 different shapes and sizes, each with its own personality. Some are water-loving, others are oil-loving, and some even have electrical charges! This incredible diversity allows proteins to fold into intricate 3D structures, like microscopic origami.

Think of it like this:

Primary structure: The basic sequence of amino acids, like a string of LEGO bricks.
Secondary structure: Simple folds and twists, like building a LEGO wall or a spiral staircase.
Tertiary structure: The complete 3D structure, like a complex LEGO castle or spaceship.
Quaternary structure: When multiple protein "castles" come together to form a mega-structure!
This intricate folding is crucial because it determines a protein's function. Even a small change in the "LEGO instructions" can have a big impact.

Amazing Protein Superstars!

Spider silk: Stronger than steel and more elastic than rubber, this protein is a true marvel of nature. Scientists are even trying to mimic its properties to create new materials for medicine, engineering, and even clothing!
Enzymes: These catalytic wizards speed up chemical reactions, making life possible. They're like tiny chefs in our cells, chopping, mixing, and cooking up the molecules we need to survive.
Collagen: The most abundant protein in our bodies, collagen is like the "glue" that holds us together. It gives our skin elasticity, strengthens our bones, and keeps our tissues connected.
Proteins are the workhorses of life, performing countless tasks that keep us healthy and thriving. From the air we breathe to the thoughts we think, proteins are involved in every aspect of our existence. So next time you marvel at the complexity of life, remember the tiny protein machines that make it all possible!

Experimental techniques and chemical analysis

Experimental design

Measurement Apparatus: A Deep Dive into Tools for Time, Temperature, Mass, and Volume

In the realm of scientific inquiry, precise measurement forms the bedrock of understanding and progress. Whether unraveling the mysteries of the cosmos or formulating new life-saving drugs, accurate measurement is paramount. This detailed exploration delves into the world of measurement apparatus, focusing on tools used for quantifying time, temperature, mass, and volume.

Time Measurement: Stopwatches

Stopwatches are indispensable tools for measuring time intervals, finding applications in diverse fields ranging from sports and athletics to scientific research and industrial processes.

They come in various forms, from traditional analog stopwatches with hands and dials to modern digital stopwatches with advanced features.

Types of Stopwatches:

Analog Stopwatches: These classic stopwatches use a mechanical mechanism with gears and springs to measure time. While they lack the precision of digital counterparts, they offer a tangible sense of time's passage.
Digital Stopwatches: These electronic stopwatches utilize quartz crystals for precise timekeeping. They typically display time in digital format, often with accuracy up to 0.01 seconds. Many digital stopwatches also include features like lap timers, split timers, and memory functions.

Applications of Stopwatches:

Sports and Athletics: Stopwatches are essential for timing races, measuring athlete performance, and analyzing training progress.
Scientific Research: Researchers use stopwatches to time reactions, monitor experiments, and collect data on time-dependent phenomena.
Industrial Processes: Stopwatches find use in manufacturing, quality control, and process optimization to measure production times and identify bottlenecks.
Case Study: In a study investigating the effects of a new drug on reaction time, researchers used digital stopwatches to precisely measure the time participants took to respond to stimuli. The stopwatches allowed for accurate data collection, enabling the researchers to draw meaningful conclusions about the drug's impact on cognitive function.

Temperature Measurement: Thermometers

Thermometers are instruments designed to measure temperature, providing insights into thermal energy and its effects on various systems. They operate on the principle that certain physical properties of materials change predictably with temperature.

Types of Thermometers:

Liquid-in-Glass Thermometers: These traditional thermometers consist of a glass tube filled with a liquid (usually mercury or alcohol) that expands or contracts with temperature changes. The liquid level against a calibrated scale indicates the temperature.
Digital Thermometers: These electronic thermometers use sensors like thermistors or thermocouples to measure temperature. They display the reading digitally, often with high accuracy and fast response times.
Infrared Thermometers: These non-contact thermometers measure temperature by detecting infrared radiation emitted by an object. They are useful for measuring temperature in situations where direct contact is not feasible or desirable.

Applications of Thermometers:

Healthcare: Thermometers are crucial for monitoring body temperature, diagnosing fevers, and tracking patient health.
Scientific Research: Thermometers are used in laboratories to control reaction temperatures, study thermal properties of materials, and conduct experiments involving temperature-sensitive processes.
Industrial Processes: Thermometers play a vital role in monitoring and controlling temperatures in manufacturing, food processing, and other industrial settings.
Case Study: In a food processing plant, infrared thermometers are used to monitor the temperature of food products during cooking and cooling processes. This ensures that the food is cooked to the correct temperature for safety and quality, while also preventing overheating or spoilage.

Mass Measurement: Balances

Balances are instruments used to determine the mass of an object, providing a quantitative measure of the amount of matter it contains. They operate on the principle of comparing the unknown mass with a known mass.

Types of Balances:

Beam Balances: These traditional balances consist of a beam with two pans suspended from its ends. The unknown mass is placed on one pan, and known masses are added to the other pan until the beam is balanced.
Electronic Balances: These modern balances use electromagnetic force restoration to measure mass. They provide highly accurate and rapid measurements, often with digital displays and features like tare function and multiple units.
Analytical Balances: These specialized balances offer extremely high precision, typically measuring mass to four or more decimal places. They are used in analytical chemistry, pharmaceutical research, and other applications requiring precise mass determination.

Applications of Balances:

Scientific Research: Balances are essential for preparing solutions, measuring reactants, and analyzing samples in chemistry, biology, and other scientific disciplines.
Industrial Processes: Balances find use in quality control, formulation, and production processes in various industries, including pharmaceuticals, food processing, and manufacturing.
Retail and Commerce: Balances are used in stores and markets to weigh goods for sale, ensuring fair trade and accurate pricing.
Case Study: In a pharmaceutical research laboratory, analytical balances are used to precisely measure the mass of drug compounds during formulation and development. This ensures accurate dosing and consistent drug efficacy.

Volume Measurement: Burettes, Pipettes, Measuring Cylinders, and Gas Syringes

Volume refers to the amount of space occupied by a substance, typically a liquid or gas. Several instruments are used to measure volume, each with its own level of accuracy and specific applications.

Burettes:

Description: Burettes are long, graduated glass tubes with a stopcock at the bottom, allowing for controlled dispensing of liquids.
Applications: Burettes are primarily used in titrations, where precise volumes of a solution are added to another solution until a chemical reaction is complete. They are also used for dispensing accurate volumes of reagents in other chemical procedures.

Volumetric Pipettes:

Description: Volumetric pipettes are designed to deliver a single, fixed volume of liquid accurately. They have a bulb-like shape with a narrow neck and a calibration mark indicating the specific volume.
Applications: Volumetric pipettes are used for transferring precise volumes of solutions in analytical chemistry, biochemistry, and other laboratory settings.

Measuring Cylinders:

Description: Measuring cylinders are graduated glass or plastic cylinders used for measuring approximate volumes of liquids. They are less accurate than burettes and pipettes but offer greater convenience for general volume measurements.
Applications: Measuring cylinders are commonly used in laboratories for dispensing reagents, preparing solutions, and measuring volumes in experiments where high accuracy is not critical.

Gas Syringes:

Description: Gas syringes are specialized syringes used to measure the volume of gases. They consist of a graduated glass or plastic barrel with a tightly fitting plunger.
Applications: Gas syringes are used in chemistry experiments to collect and measure gases produced in reactions, as well as to deliver precise volumes of gases for reactions or analysis.
Case Study: In a water treatment plant, measuring cylinders are used to measure the volume of chemicals added to the water for purification. This ensures that the correct amount of chemicals is used to effectively treat the water without overdosing.

Conclusion

The accurate measurement of time, temperature, mass, and volume is fundamental to scientific inquiry and technological advancement. The tools discussed in this exploration – stopwatches, thermometers, balances, burettes, pipettes, measuring cylinders, and gas syringes – represent a diverse array of instruments designed for precise and reliable measurement. As technology continues to evolve, we can expect further advancements in measurement apparatus, leading to even greater accuracy and new possibilities for scientific discovery and innovation.

Unveiling the Magic and Mayhem of Scientific Inquiry (2024 Update)

Imagine yourself as a detective, piecing together the puzzle of the universe. Your tools? Experimental methods and apparatus, each with their own quirks and charms. But like any good detective story, there's always a twist! Let's dive into this thrilling world of scientific exploration, where we'll uncover the strengths, weaknesses, and hidden surprises of these investigative tools.

I. Experimental Methods: A Detective's Toolkit

A. The Classics: Tried and True

Controlled Experiments: The Sherlock Holmes Approach
Think of controlled experiments as the Sherlock Holmes of research methods. Meticulous, precise, and always in control. By isolating variables like a master detective isolates clues, we can pinpoint cause and effect with incredible accuracy.

The Good: Unraveling mysteries with laser focus, ensuring rock-solid evidence, and replicating findings with ease.

The Not-So-Good: Sometimes the real world is messier than our controlled environments. Ethical dilemmas can arise, and what we learn in the lab might not always hold true in the wild.

Example: Imagine testing a new super-strength serum. You'd give it to one group of volunteers (your superheroes-in-training) and a placebo to another (the control group). Then, you'd unleash them on an obstacle course, carefully monitoring their performance while keeping everything else (diet, sleep, background music) the same.

Natural Experiments: Nature's Grand Experiment

Sometimes, the universe sets up the experiment for us. Natural experiments are like stumbling upon a crime scene already in progress. We observe the consequences of natural events, like a volcanic eruption or a sudden economic shift, to understand their impact.

The Good: Real-world insights, tackling ethical challenges head-on, and the thrill of studying rare events.

The Not-So-Good: Lack of control can muddy the waters, making it tricky to pinpoint exact causes. Plus, these opportunities can be rare and difficult to replicate.

Example: Imagine a remote island where, by chance, half the population starts eating a unique fruit. Years later, you discover they have incredible longevity. A natural experiment in action!

Field Experiments: The Undercover Operation

Field experiments are the undercover agents of the research world. They bring the control of a lab experiment into the real world, blending in seamlessly.

The Good: A perfect balance of control and realism, capturing behavior in its natural habitat.

The Not-So-Good: Ethical considerations become trickier when people are unaware, they're part of an experiment. Control is still limited compared to the lab, and every field setting is unique, making replication a challenge.

Example: Imagine subtly changing the music in a store to see if it influences buying behavior. That's a field experiment in action, subtly manipulating the environment to understand its effects.

B. The New Kids on the Block: Cutting-Edge Tools

Virtual Reality (VR) Experiments: Stepping into the Matrix
VR technology is like stepping into the Matrix. We can create immersive, interactive worlds to study human behavior in ways never before imagined.

The Good: Unprecedented control over the environment, pushing the boundaries of realism, and exploring scenarios impossible to create in the real world.

The Not-So-Good: VR is still evolving, with limitations in technology and potential ethical concerns around participant safety and data privacy.

Example: Imagine treating phobias by exposing people to their fears in a safe, controlled VR environment. Or training surgeons on complex procedures without any real-world risks.

Online Experiments: The Global Laboratory

The internet has transformed our world, and research is no exception. Online experiments tap into the vast, diverse online population, opening up a world of possibilities.

The Good: Massive sample sizes, cost-effectiveness, and global reach.

The Not-So-Good: Ensuring data quality and battling potential biases in online samples can be tricky.

Example: Think of A/B testing different website layouts to see which one keeps users engaged longer. That's the power of online experiments.

Big Data and Machine Learning: The Data Detectives

Big data and machine learning are like having a team of super-powered analysts sifting through mountains of information. They can uncover hidden patterns and insights that would be impossible for humans to find alone.

The Good: Unmatched precision, personalized interventions, and predictive power.

The Not-So-Good: Data bias can be amplified, and understanding the complex results can be a challenge. Ethical considerations around privacy and algorithmic bias are crucial.

Example: Imagine analyzing massive datasets of student performance to identify the factors that contribute to success and create personalized learning plans.

II. Experimental Apparatus: The Detective's Gadgets

Just like a detective needs the right tools for the job, scientists rely on a fascinating array of apparatus. Choosing the right gadget can make or break an experiment!

A. Choosing the Right Tool for the Job

Factors to consider include:

Precision and Accuracy: Our tools need to be as sharp as our minds!
Sensitivity: Can it detect the subtlest of changes?
Range: Is it suitable for the scale of our investigation?
Reliability: Can we trust it to deliver consistent results?
Cost and Availability: Even the best detectives have budgets!
Ease of Use: We need tools that are user-friendly, not head-scratchers.
Safety: Safety first! No exploding test tubes, please.

B. Gadgets Galore: A Glimpse into the Lab

Microscopes: The ultimate zoom lens, revealing the hidden wonders of the microscopic world.
Spectrophotometers: Analyzing the interaction of light and matter, unlocking the secrets hidden within substances.
Chromatography Systems: Separating mixtures like a master chef, isolating individual components for analysis.
Electrophysiological Equipment: Listening in on the electrical conversations of the body, from brainwaves to muscle activity.
Sensors and Data Loggers: The silent observers, tirelessly collecting data while we sleep.
Each of these tools has its own strengths and weaknesses, and choosing the right one is crucial for a successful investigation.

A Quirky Chemist's Guide to Solutions, Solutes, and Solvents

Ever wondered how your morning coffee dissolves sugar so perfectly? Or why oil and water just refuse to mix? Well, my friend, you've stumbled into the fascinating world of solutions, solutes, and solvents!

Think of it like this:

Imagine you're throwing a party (a solution!). You've got your main ingredient, the life of the party, which is your solvent. Let's say it's a big bowl of punch. Then you've got your guests, the solutes, who mingle and mix into the punch.

The Solvent: The Party Host

Personality matters: Some solvents are polar (like water), meaning they have a positive and negative side, like a magnet. Others are nonpolar (like oil), and they're a bit more laid-back. "Like dissolves like," so polar solvents hang with polar solutes, and nonpolar solvents stick with their own kind. That's why water loves salt but hates oil!
The Mood Setter: Some solvents are volatile, meaning they evaporate quickly, like a party guest who disappears in a puff of smoke (think acetone!). Others are more chill, like water, and stick around longer.
Thick or Thin? Viscosity is all about how thick a solvent is. Think honey (high viscosity) versus water (low viscosity).

The Solute: The Party Guests

Different Flavors: Solutes can be solids (like sugar), liquids (like food coloring), or even gases (like the CO_2 that makes your soda fizzy).
Mingling Manners: How well a solute dissolve depends on the solvent, the temperature, and even the pressure (especially for gases). Think of it like how some guests mingle easily, while others prefer to stick to their own group.

The Solution: The Party Itself

A Happy Mix: A solution is a homogenous mixture, meaning everything is evenly spread out, like a good party where everyone's having fun!
Types of Parties: You can have solid solutions (like alloys in jewelry), liquid solutions (like your coffee), and even gaseous solutions (like the air we breathe!).

Saturated Solution: The Party's Full

Guest Limit Reached: A saturated solution is like a party that's reached its maximum capacity. No more guests (solutes) can fit in!
Dynamic Dance Floor: Even in a saturated solution, there's still a lot of movement. Solutes are constantly dissolving and "un-dissolving," like guests moving on and off a crowded dance floor.

Residue: The Leftovers

Party's Over: After the party, you've got some leftovers (residue). This could be anything from the crumbs on the floor to the melted ice in the punch bowl.

Filtrate: The Strained Punch

Cleaning Up: If you strain your punch to get rid of the fruit pulp, the clear liquid you get is the filtrate.
So, there you have it! A quirky chemist's guide to solutions, solutes, and solvents. Next time you're mixing up a drink or watching your bath bomb fizz, you'll have a whole new appreciation for the chemistry happening right before your eyes!

A Quirky Chemist's Guide to Acid-Base Titration

Imagine a battle between two powerful forces: the acidic army, led by the sour-faced Hydrogen ions (H+), and the basic brigade, commanded by the slippery Hydroxide ions (OH-). They clash in a swirling vortex of chemical warfare, vying for dominance. But wait! A mysterious figure emerges, the "Indicator," a shape-shifting spy who changes color to signal the end of this epic struggle. This, my friends, is the world of acid-base titration!

1. The Titration Toolkit & Battle Plan

Every chemist needs their trusty tools for this chemical showdown:

Burette: A long, slender weapon, like a magic wand, precisely dispensing the "titrant" (our champion solution) drop by glorious drop.
Pipette: A precise measuring device, like a tiny ladle, scooping up the "analyte" (the mystery solution) and placing it in the battle arena.
Titration Flask: The grand battlefield, a conical flask where the acidic and basic armies clash.
Indicator: The cunning spy, a substance that dramatically changes color when the battle reaches its peak, signaling the "equivalence point" – the moment of perfect balance between acids and bases.
Stand and Clamp: A sturdy support system to hold the burette aloft, ensuring a fair fight.
Wash Bottle: A cleansing fountain of distilled water, rinsing the battlefield and weapons to maintain the purity of the contest.

The Battle Plan:

Prepare the Analyte: Carefully measure the analyte with the pipette and transfer it to the titration flask. Add a few drops of the indicator – the spy is in position!
Prepare the Titrant: Fill the burette with the titrant, our known champion, after a thorough rinse. Note the initial volume – the battle is about to begin!

Titration Commence! Slowly release the titrant from the burette into the flask, swirling the mixture like a whirling dervish to ensure a complete reaction.
Victory is Near! Watch closely for the indicator's dramatic color change – the signal that the equivalence point has been reached. Record the final titrant volume.
Calculate the Spoils of War: Use the magical formula: Concentration of analyte = (Volume of titrant * Concentration of titrant) / Volume of analyte

2. The Shape-Shifting Spies: Indicators

These cunning spies are weak acids or bases that transform their appearance based on the surrounding environment. Some famous spies include:

Phenolphthalein: Colorless in acidic realms, it blushes a vibrant pink in the presence of a base. A true drama queens!
Methyl Orange: Red-faced in acidic conditions, it turns a cheerful yellow when surrounded by a base. A mood ring of the chemical world!
Bromothymol Blue: A triple agent! Yellow in acidic territory, blue in basic lands, and a tranquil green at the neutral point. A true master of disguise.

3. The End-Point: Declaring the Victor

The moment the indicator changes color is the "end-point" – the apparent victory. But beware! It might not perfectly coincide with the equivalence point, the true moment of balance. This discrepancy is the "titration error," a sneaky trickster! To minimize this error, choose your indicator wisely and add the titrant drop by agonizing drop near the end-point, like a skilled archer aiming for the bullseye.

4. Types of Acid-Base Battles

The chemical world hosts various types of acid-base battles:

Strong Acid vs. Strong Base: A classic clash of titans, resulting in a neutral battlefield (pH 7).

Strong Acid vs. Weak Base: The strong acid dominates, leaving the battlefield slightly acidic ($pH < 7$).
Weak Acid vs. Strong Base: The strong base takes charge, making the battlefield slightly basic ($pH > 7$).
Weak Acid vs. Weak Base: A subtle and complex struggle, where the equivalence point is elusive and difficult to pinpoint.

5. Applications: Where the Battles Matter

These chemical battles aren't just for show; they have real-world applications!

Chemical Analysis: Unmasking the true identity and strength of acids and bases in various concoctions, from industrial chemicals to life-saving drugs.
Environmental Monitoring: Assessing the health of our planet by measuring the acidity or alkalinity of water and soil.
Biological Research: Delving into the secrets of life by analyzing bodily fluids like blood and urine.
Industrial Processes: Maintaining a delicate balance in manufacturing processes, ensuring the quality of products like fertilizers and medicines.

6. Case Studies: Tales from the Lab

The Vinegar Mystery: Unraveling the secrets of vinegar by battling its acetic acid content with a valiant sodium hydroxide solution, guided by the ever-reliable phenolphthalein.
The Water Puzzle: Deciphering the alkalinity of water – its ability to neutralize acids – by pitting it against a standard hydrochloric acid solution, with methyl orange as the watchful spy.

7. The Future of Chemical Warfare: Advanced Titration Techniques

The world of titration is constantly evolving:

Automated Titration Systems: Robotic warriors that perform titrations with incredible precision and speed, leaving no room for human error.
Microfluidic Devices: Tiny battlefields, allowing for analysis of minuscule samples, pushing the boundaries of chemical investigation.

8. Epilogue: The Never-Ending Quest

Acid-base titration is a cornerstone of chemistry, a captivating dance between acids and bases, guided by the wisdom of indicators. This quirky guide has hopefully shed light on this fundamental technique, its myriad applications, and its exciting future. So, embrace the world of chemistry, explore its mysteries, and remember – the battle between acids and bases is a never-ending quest for knowledge and understanding!

Imagine a bustling city: streets teeming with diverse people, each with their own unique destination and pace. Paper chromatography is like this city, where molecules, instead of people, navigate through a network of cellulose fibers.

The Paper City:

Buildings (Cellulose Fibers): These form the intricate cityscape, providing pathways and obstacles for our molecular travelers.
Roads (Solvent): A flowing river of solvent acts as the main thoroughfare, carrying molecules along its course.
Travelers (Molecules): Each molecule, with its unique size, shape, and personality (polarity), embarks on a journey through the city.
The Journey:

Some molecules, drawn to the allure of the buildings, linger and explore the shops and cafes (cellulose fibers). Others, eager to reach their destination, rush along the flowing river (solvent). This difference in pace and attraction separates the travelers, creating distinct groups along the cityscape.

Unveiling the Hidden:

Imagine some travelers are invisible. To reveal their presence, we use special "detectives" (locating agents) who can make them visible. These detectives might shine a special light (UV lamp) or use their sense of smell (chemical reagents) to uncover the hidden travelers.

The Applications:

This bustling city of molecules has many uses:

Forensic Science: Detectives use it to analyze ink from a ransom note or identify traces of poison.
Food Industry: Inspectors ensure the quality of our food by identifying artificial colors or detecting contaminants.
Medicine: Scientists separate and analyze components of drugs to ensure their safety and effectiveness.

Advancements:

The city is constantly evolving:

Expressways (HPPC): New, faster routes are being developed for quicker and more efficient travel.
Multi-level City (Two-Dimensional Chromatography): The city expands vertically, offering more complex routes for separating even the most similar travelers.
Paper chromatography, like a bustling city, is a dynamic and versatile technique. It allows us to separate, identify, and analyze the diverse molecules that make up our world.

Unlocking the Secrets of Chromatograms: A Colorful Journey into Mixture Mysteries (2024)

Imagine a detective meticulously separating clues to solve a complex case. That's precisely what chromatography does! It's a scientific technique that acts like a molecular detective, separating the individual components hidden within a mixture. Think of it as an intricate dance where different molecules, with their unique personalities, move at varying speeds through a special obstacle course. The result? A vibrant and insightful chromatogram, a visual record of this molecular dance-off.

Delving into the Chromatographic Maze

This guide is your passport to the fascinating world of chromatograms. We'll embark on a journey to decipher these colorful patterns, learning how to identify mysterious substances and distinguish between the pure and the impure. Our adventure will cover:

Chromatography Unveiled: We'll unravel the magic behind this separation process, understanding how molecules interact with their surroundings.
Retention Time: The Molecular Stopwatch: Discover how the time it takes for a molecule to navigate the course reveals its identity.
Peak Shape and Size: The Telltale Signs: We'll learn how the shape and size of the peaks in a chromatogram provide clues about the nature and amount of each component.

Chromatogram Comparison: A Molecular Lineup: Just like detectives compare fingerprints, we'll compare chromatograms to identify unknown substances by matching them with known culprits.
Purity Assessment: Separating the Wheat from the Chaff: We'll explore how to determine the purity of a substance based on its chromatographic fingerprint.
Real-World Applications: Chromatography in Action: From catching criminals to ensuring the safety of our food and medicines, we'll see how chromatography plays a vital role in our everyday lives.

Types of Chromatography: A Diverse Toolkit

Just like a detective uses different tools for different investigations, scientists use various types of chromatography, each with its own strengths and specialties. We'll encounter:

Gas Chromatography (GC): The high-speed chase for volatile molecules that love to fly.
High-Performance Liquid Chromatography (HPLC): The elegant waltz for non-volatile molecules that prefer a more leisurely pace.
Thin-Layer Chromatography (TLC): The quick and versatile sketch artist, perfect for initial investigations.
Paper Chromatography: The classic technique, often used to introduce budding scientists to the wonders of separation.

Retention Time (Rt): The Molecular Stopwatch

Imagine each molecule having its own internal clock that starts ticking the moment it enters the chromatographic race. The time it takes to reach the finish line is its retention time (Rt). This is a crucial clue in our molecular detective work, as it helps us identify different substances.

Peak Shape and Size: The Descriptive Profile

The peaks in a chromatogram are not just colorful blobs; they hold valuable information about the character of each component.

Peak Shape: A sharp, symmetrical peak suggests a well-behaved molecule, while a tailing or fronting peak hints at some intriguing interactions with the surrounding environment.
Peak Size: The area or height of a peak tells us how much of a particular substance is present in the mixture.

Chromatogram Comparison: The Molecular Lineup

By comparing the chromatogram of an unknown substance to those of known standards, we can identify the mystery molecule. It's like a police lineup, where we try to match the suspect's fingerprint with those in the database.

Purity Assessment: Unveiling Hidden Impurities

A pure substance should ideally produce a single, sharp peak in the chromatogram. Any extra peaks or distortions are telltale signs of impurities lurking within.

Applications and Case Studies: Chromatography in the Real World

Chromatography is not just a theoretical exercise; it has a profound impact on our lives. We'll explore how it's used in:

Pharmaceutical Industry: Ensuring the purity and safety of medicines.
Environmental Monitoring: Detecting pollutants in our air and water.
Food Industry: Analyzing food composition and ensuring its safety.
Forensic Science: Helping to solve crimes by analyzing evidence.
Beyond the Basics: Advanced Chromatographic Adventures

For those eager to delve deeper, we'll touch upon advanced techniques like peak deconvolution, mass spectrometry, and chemometrics, which allow us to extract even more information from these molecular fingerprints.

Conclusion: Embracing the Chromatographic Challenge

Chromatogram interpretation is like learning a new language, one that speaks volumes about the hidden world of molecules. By mastering this language, we can unlock the secrets of mixtures, identify unknown substances, and ensure the quality and safety of products that impact our lives. So, join us on this colorful journey of discovery, and let's unravel the mysteries hidden within chromatograms together!

Rf Value in Chromatography: Unmasking the Mystery of Mixtures

Imagine a bustling city with diverse individuals moving at different speeds. Some are drawn to window shopping, while others rush towards their destinations. Chromatography is like observing this city from above, where we separate the "individuals" (components) of a "crowd" (mixture) based on their unique "walking styles" (affinities to different phases).

In this bustling city of molecules, the Rf value is our detective tool. It's a special "speedometer" that tells us how fast each component moves compared to the fastest one.

Calculating the "Speedometer" Reading:

Think of a race where the solvent is the leader, setting the pace. The Rf value is simply the ratio of the distance traveled by a component to the distance covered by the solvent front. It's like saying, "This molecule traveled half the distance the solvent did."

Why This "Speedometer" Matters:

Unmasking Hidden Identities: Each molecule has a characteristic Rf value under specific conditions. It's like their unique fingerprint, helping us identify them in a crowd.
Spotting the Imposters: A pure substance will have a single, well-defined spot, like a solo traveler with a clear purpose. Multiple spots or streaks reveal impurities, like a group trying to blend in.
Fine-tuning the "Traffic Flow": Rf values help us optimize the separation process. If the components are moving too slowly, we can give them a "push" by changing the solvent.

Factors Influencing the "Speedometer":

Just like traffic conditions affect travel time, several factors influence Rf values:

The Road: The nature of the stationary phase is like the road surface – some are smooth, others rough, affecting the molecule's journey.
The Vehicle: The mobile phase is like the vehicle carrying the molecules – a powerful car speeds things up, while a bicycle takes longer.
The Weather: Temperature affects the energy of the molecules, like warm weather encouraging faster movement.
The Crowd: Overloading the sample is like overcrowding the city, leading to confusion and inaccurate readings.

Real-life Adventures with Rf Values:

Identifying Amino Acids: Imagine separating a protein shake into its individual amino acids, like identifying each ingredient in a complex recipe.
Purity Assessment of a Drug: Ensuring a medicine is pure is like checking for contaminants in your food – we want the real deal, not unwanted extras.
Forensic Analysis: Rf values help forensic scientists analyze evidence, like identifying the ink used in a forged document or the drugs present in a sample.

Beyond the Basics:

Relative Rf Value: To ensure accuracy, we often compare the Rf value of an unknown component with a known standard, like comparing your speed to a pace car.
Two-Dimensional Chromatography: This is like exploring the city from two different angles, providing a more comprehensive view of the components.
High-Performance Liquid Chromatography (HPLC): Imagine a high-speed train for molecules, offering faster and more efficient separation.

In Conclusion:

The Rf value is a powerful tool in the world of chromatography, helping us unravel the mysteries of mixtures. By understanding its significance and the factors that influence it, we can unlock valuable insights into the composition and purity of substances. So, next time you encounter a mixture, remember the bustling city of molecules and the detective work of Rf values!

Methods of Separation and Purification: A Chemist's Toolkit

Imagine a world where it was impossible to separate the good from the bad, the pure from the impure. Chaos would resign! Luckily, chemists have a whole arsenal of tools for just that purpose: separating and purifying substances. Think of it like sorting Legos, but at a molecular level. Let's dive into some of the most common techniques:

(a) Solvent Extraction: Like Dissolves Like

Ever made salad dressing? You know how oil and vinegar separate? That's the basic idea behind solvent extraction. We exploit the fact that some things dissolve better in one liquid than another. It's like choosing the right fishing lure for the fish you want to catch.

The principle: "Like dissolves like." Polar compounds (like salt) dissolve in polar solvents (like water), while nonpolar compounds (like oil) dissolve in nonpolar solvents (like hexane).
The Process: Mix your mixture with the chosen solvent, shake it up, and let it settle. The compound you want will move into the solvent, while the rest stays put. Then, just evaporate the solvent, and voila! You've got your purified substance.

Real-world examples:
Coffee: Caffeine is extracted from coffee beans using dichloromethane.
Perfumes: Fragrant essential oils are extracted from plants using solvents like ethanol.
Medicine: The antimalarial drug artemisinin is extracted from sweet wormwood using hexane.

(b) Filtration: A Molecular Sieve

Think of a coffee filter. It lets the liquid through but holds back the grounds. Filtration in chemistry is similar. We use a porous material to separate solids from liquids or gases.

The principle: Particle size matters! The filter has tiny holes that let small molecules (like water) pass through while trapping larger ones (like sand).

Types of Filtrations:
Gravity Filtration: Simple and slow, like making drip coffee.
Vacuum Filtration: Faster, like using a shop vac.
Hot Filtration: For substances that solidify easily, we keep them warm during filtration.
Real-world examples:
Water purification: Filters remove dirt, bacteria, and other nasties from our drinking water.
Chemistry labs: Separating solid byproducts from reaction mixtures.

(c) Crystallization: Building Molecular Jewels

Have you ever seen rock candy? That's sugar that's been crystallized. Crystallization is a beautiful way to purify solid compounds.

The principle: Most solids dissolve better in hot liquids than cold ones. As a hot solution cools, the compound becomes less soluble and starts to form crystals, leaving impurities behind in the liquid.
The Process: Dissolve the impure solid in a hot solvent, filter out any insoluble junk, then slowly cool the solution. Crystals will form, which can then be separated and dried.
Real-world examples:
Sugar production: Sugar crystals are formed from sugarcane or beet juice.
Pharmaceuticals: Many drugs are purified by crystallization.
Chemistry experiments: Purifying benzoic acid (a common food preservative).

(d) Simple Distillation: Separating by Boiling Point

Imagine you have a mixture of salt and water. How do you separate them? Simple distillation! We heat the mixture, and the water boils off first, leaving the salt behind.

The principle: Different liquids have different boiling points. The one with the lower boiling point will vaporize first.

The Process: Heat the mixture, the liquid with the lower boiling point turns to vapor, the vapor is cooled and condenses back into a liquid in a separate container.
Real-world examples:
Desalination: Removing salt from seawater to make freshwater.
Purifying solvents: Like ethanol or acetone.

(e) Fractional Distillation: Fine-Tuning the Separation

What if you have liquids with similar boiling points? That's where fractional distillation comes in. It's like simple distillation, but with an extra step to improve the separation.

The principle: Still based on boiling points, but we use a fractionating column to create multiple "mini-distillations," allowing for finer separation of liquids with close boiling points.

Real-world examples:
Crude oil refining: Separating gasoline, kerosene, and other components from crude oil.
Producing alcoholic beverages: Separating ethanol from water to make spirits.
These are just a few of the many technique's chemists use to separate and purify substances. It's a fascinating world of molecular manipulation, and it plays a vital role in everything from medicine and food production to environmental science and manufacturing.

Separation Anxiety: A Chemistry Love Story (2024 Edition)
Forget Romeo and Juliet! In the dramatic world of chemistry, molecules face their own passionate struggles. Their ultimate goal? To be together, yet apart. Join us as we unravel the epic tales of separation and purification, where compounds embark on journeys to find their true, unadulterated selves.

(Cue dramatic music and a swirling vortex of colorful liquids)

Part 1: The Quest for Purity - Breaking Up is Hard to Do

Imagine a bustling dance floor packed with molecules, all bumping and grinding in a chaotic mix. But some molecules yearn for individuality, craving escape from the mosh pit of impurities. That's where our separation heroes' step in, each with their unique charm and strategy:

Filtration: The Wallflower: Like a shy introvert at a party, filtration prefers to hang back and observe. Using a trusty filter paper as its wingman, it patiently separates the solid wallflowers from the liquid social butterflies, based on their size and social awkwardness (particle size, that is).

Distillation: The Drama Queen: This diva demands attention, boiling over with excitement at the slightest provocation. With a flair for the dramatic, distillation exploits boiling point differences, leaving behind the less volatile, more grounded molecules.

Crystallization: The Ice Queen: Patience is her virtue, elegance her weapon. Crystallization seductively lures a compound into a hot embrace (solvent), only to slowly withdraw her affection, leaving behind stunning, pure crystals as a cold reminder of her power.

Chromatography: The Social Butterfly: This technique thrives in a crowd, effortlessly flitting between different phases. With an uncanny ability to play the field, chromatography separates molecules based on their unique social connections (affinities) to different materials.

Extraction: The Homewrecker: A master of seduction, extraction entices a molecule away from its current partner (mixture) with the promise of a more compatible soulmate (solvent). Two become three in this love triangle, leading to a clean break and a fresh start.

Sublimation: The Ghost: Some molecules are just too cool for the liquid state, preferring to vanish into thin air (gas phase) and reappear in a purified form. Sublimation, the master of disappearing acts, helps these elusive compounds bypass the drama of melting altogether.

Centrifugation: The Whirlwind: This technique takes matters into its own hands, literally spinning the mixture into a frenzy. Density becomes the deciding factor as heavier molecules are flung to the edges, leaving the lighter ones in the center stage.

Electrophoresis: The Electric Slide: In this electrifying separation, charged molecules are drawn to their opposite partners, creating a dazzling dance of attraction and repulsion.

(Imagine a montage of molecules undergoing these techniques, set to a catchy pop song about breaking free)

Choosing the Right Move:

Just like in any relationship, choosing the right separation technique requires careful consideration. Is it a casual fling (simple mixture) or a complex entanglement (complex mixture)? Are the partners volatile and dramatic, or cool and collected? The key is to find the technique that best suits the personalities involved.

Part 2: Melting Point & Boiling Point - The Love Test

In the quest for true love (purity), melting point (MP) and boiling point (BP) are the ultimate tests of a compound's character. A sharp MP, like a passionate kiss, signifies purity, while a broad MP, like a lukewarm handshake, hints at lingering impurities. BP, on the other hand, reveals a compound's volatility and staying power.

(Imagine a romantic scene with two molecules gazing into each other's eyes, a thermometer between them)

Case Studies: Love in Action

Aspirin's Purification Journey: Poor aspirin, synthesized from a messy reaction, is burdened with impurities. But fear not! Crystallization comes to the rescue, dissolving aspirin in a warm embrace of ethanol and then slowly cooling the relationship, leaving behind pure, sparkling crystals.

Caffeine's Escape from Tea Leaves: Caffeine, trapped within the confines of tea leaves, yearns for freedom. Extraction and sublimation join forces to liberate this stimulating compound, first by enticing it with hot water and then whisking it away into the gas phase, leaving behind the mundane impurities.

Drug Mixture's Identity Crisis: A mysterious drug mixture needs to be unmasked. HPLC, the ultimate detective, steps in, separating the components based on their unique personalities and revealing their true identities.

(Imagine a detective examining a chromatogram with a magnifying glass)

Epilogue:

In the grand tapestry of chemistry, separation and purification are the unsung heroes, ensuring that compounds can express their true selves. By understanding their unique personalities and employing the right techniques, we can unlock the secrets of the molecular world and create a more harmonious and pure existence.

Unmasking the Invisible: A Detective's Guide to Anion Identification

Imagine yourself as a chemical detective, tasked with uncovering the hidden identities of mysterious suspects – anions, the negatively charged ions lurking in unknown solutions. Armed with your trusty toolkit of reagents and a keen eye for observation, you're ready to embark on a thrilling quest to unmask these elusive culprits.

1. The Case of the Fizzy Fugitive: Carbonate (CO_3^{2-})

Our first suspect is known for its dramatic escape attempts. Add a few drops of dilute hydrochloric acid (HCl) to the solution, and if carbonate is present, it'll make a break for it as a colorless, odorless gas – carbon dioxide (CO_2). To confirm its identity, pass the escaping gas through limewater (calcium hydroxide solution). If the limewater turns cloudy, you've caught your carbonate red-handed, trapped in the form of a white precipitate of calcium carbonate ($CaCO_3$).

Think of it like this: Imagine dropping a fizzy tablet into water. The bubbles that rise are carbon dioxide, just like in our carbonate test.

Real-world application: Geologists use this test to identify carbonate rocks like limestone. A drop of acid on the rock will cause it to fizz if carbonate is present.

2. The Halide Trio: Chloride (Cl^-), Bromide (Br^-), and Iodide (I^-)

These three suspects are close relatives, often found mingling together. To tell them apart, we'll use a silver nitrate solution. Add a few drops, and watch for the formation of a precipitate. The color of the precipitate is your clue:

Chloride (Cl^-): A white, curdled precipitate like milk.
Bromide (Br^-): A creamy precipitate, the color of pale butter.
Iodide (I^-): A vibrant yellow precipitate, like a sunny day.
Think of it like this: Imagine developing an old photograph. The silver compounds used in photography are similar to the precipitates formed in this test.

Real-world application: Forensic scientists use this test to analyze gunshot residue, helping to link a suspect to a crime scene.

3. The Pungent Phantom: Nitrate (NO_3^-)

This suspect is known for its strong, unmistakable odor. To unmask it, add a piece of aluminum foil and a few drops of sodium hydroxide to the solution, then gently heat it. If nitrate is present, it'll transform into ammonia gas (NH_3), with its characteristic pungent smell. To be absolutely sure, hold a damp red litmus paper near the test tube. If it turns blue, you've caught your nitrate.

Think of it like this: Imagine the strong smell of smelling salts. That's ammonia, the same gas produced in our nitrate test.

Real-world application: Environmental scientists use this test to monitor nitrate levels in water, which can indicate pollution from fertilizers.

4. The Cloudy Conspirator: Sulfate (SO_4^{2-})

This suspect prefers to stay hidden in plain sight. To reveal its presence, add a few drops of barium nitrate solution. If sulfate is present, it'll form a dense white precipitate of barium sulfate ($BaSO_4$), clouding the solution.

Think of it like this: Imagine adding milk to water. The milk particles disperse and make the water cloudy, similar to how barium sulfate makes the solution cloudy.

Real-world application: In the construction industry, this test helps determine the sulfate content in cement, which can affect the durability of concrete structures.

5. The Chameleon: Sulfite (SO_3^{2-})

This master of disguise is known for its ability to change colors. To expose its true identity, add the test solution to a purple potassium permanganate solution. If sulfite is present, it'll steal the color from the permanganate, turning it colorless.

Think of it like this: Imagine a chameleon blending into its surroundings. The sulfite, like a chameleon, changes the color of the solution to conceal itself.

Real-world application: Winemakers use this test to measure sulfite levels in wine, which are used as preservatives.

Conclusion

With your newfound knowledge and detective skills, you're now equipped to identify these common anions. Remember, the world of chemistry is full of mysteries waiting to be solved, and with careful observation and experimentation, you can unlock the secrets hidden within.

Unmasking the Mysterious Aqueous Cations: A Detective's Guide to Sodium Hydroxide and Ammonia

Imagine yourself as a chemical detective, faced with a lineup of mysterious aqueous cations. Your mission: to unmask their true identities using only two trusty sidekicks – sodium hydroxide (NaOH) and ammonia (NH3). Intrigued? Let's dive into this captivating world of colorful reactions and clever deductions!

The Science of Sleuthing

Our detective work hinges on the unique reactions of different cations with the hydroxide ions (OH-) unleashed by NaOH and NH3. These reactions often lead to the formation of eye-catching precipitates – insoluble metal hydroxides with distinct colors and textures – or complex ions that whisper secrets about their central metal cation.

Rogues Gallery: Cation Profiles

Aluminum (Al3+): This slippery character initially forms a white, gelatinous disguise (Al (OH)3) with both NaOH and NH3. However, excess NaOH dissolves this mask, revealing a soluble tetra hydro xo -

aluminate (III) complex ion. NH3, on the other hand, sees through this trickery, leaving the precipitate intact.

Ammonium (NH4+): A master of disguise, ammonium doesn't form a precipitate with NaOH. Instead, it releases a pungent ammonia gas (NH3), giving itself away with its characteristic odor and ability to turn damp red litmus paper blue. This trickster remains silent when confronted with NH3.

Calcium (Ca2+): Calcium prefers a subtle approach, forming a faint white precipitate (Ca (OH)2) with NaOH. This shy precipitate might even disappear in dilute solutions. NH3, however, fails to elicit any response.

Chromium (III) (Cr3+): This cunning cation initially dons a green, gelatinous cloak (Cr (OH)3) with both NaOH and NH3. Excess NaOH, however, persuades it to shed this disguise, revealing a green hex a hydro xo- chromate (III) complex ion. NH3 remains unimpressed, leaving the precipitate undisturbed.

Copper (II) (Cu2+): A flamboyant character, copper (II) initially forms a pale blue precipitate (Cu (OH)2) with both NaOH and NH3. However, excess NH3 coaxes it into a dramatic transformation, revealing a deep blue tetra ammine copper (II) complex ion. This striking color is a dead giveaway!

Iron (II) (Fe2+): This quick-change artist initially forms a green precipitate (Fe (OH)2) with both NaOH and NH3. However, it rapidly oxidizes in air, turning brown (Fe (OH)3) like a chameleon changing its colors.

Iron (III) (Fe3+): A bold and straightforward character, iron (III) immediately forms a reddish-brown precipitate (Fe (OH)3) with both NaOH and NH3, leaving no room for doubt about its identity.

Zinc (Zn2+): This deceptive cation initially forms a white, gelatinous disguise (Zn (OH)2) with both NaOH and NH3. However, excess NaOH or NH3 dissolves this mask, revealing a colorless tetra hydro xo- zincate (II) or tetra ammine- zinc (II) complex ion, respectively.

Case Files: Cracking the Cation Codes

Case 1: The Unknown Suspect: A single cation lurks in a mysterious solution. NaOH produces a white precipitate that dissolves in excess, while NH3 forms a persistent white precipitate. Our detective work points to aluminum (Al3+) as the culprit!

Case 2: Iron (II) or Iron (III)? A solution holds one of two iron suspects. NaOH produces a green precipitate that quickly turns brown, suggesting iron (II). NH3 confirms this suspicion with a similar color change. Alternatively, we could employ potassium hexacyanoferrate (III) (for iron (II)) or potassium hexacyanoferrate (II) (for iron (III)) to produce distinctive deep blue precipitates – Turnbull's blue and Prussian blue, respectively.

Case 3: A Cationic Duo: A solution harbors both copper (II) and zinc. NH3 produces a pale blue precipitate that dissolves in excess, revealing a deep blue solution – a clear sign of copper (II). After filtering out the copper (II) ions, NaOH added to the remaining solution produces a white precipitate that dissolves in excess, confirming the presence of zinc.

Epilogue: The End of the Chemical Mystery

By carefully observing the formation, color, and solubility of precipitates with NaOH and NH3, we've successfully unmasked the identities of our aqueous cation suspects. This detective work is not just a game; it's a fundamental skill for chemists, enabling them to understand the properties of elements and their compounds, with applications spanning various scientific fields. So, embrace your inner chemical detective and continue exploring the fascinating world of qualitative analysis!

Unlocking the Secrets of the Air: A Gas Detective's Handbook

Imagine yourself as a detective, but instead of chasing down criminals, you're on the hunt for elusive gases – invisible substances that whisper secrets about the world around us. Intrigued? Let's embark on this thrilling adventure of gas identification, where we'll unravel the mysteries hidden within seemingly empty test tubes!

Our Arsenal of Tests: Where Chemistry Meets Detective Work

Just like a detective relies on fingerprints and clues, we have our own set of ingenious tools – chemical tests – to unveil the identity of these gaseous suspects. These tests are like magic tricks, producing dramatic color changes, cloudy concoctions, and even fiery bursts that reveal the true nature of each gas.

(a) Ammonia (NH_3): The Case of the Turning Tide

Ammonia, a pungent gas with a nose-wrinkling reputation, is our first case. We'll use a simple slip of red litmus paper, dampened with water, as our secret weapon. If the paper turns blue, we've caught ammonia red-handed! It's like witnessing a tide suddenly changing direction, revealing the hidden presence of this alkaline gas.

Real-world applications: Think of sniffing out ammonia leaks in factories, ensuring our drinking water is safe, or even diagnosing illnesses with a simple breath test.

(b) Carbon Dioxide (CO_2): The Cloudy Caper

Next up is carbon dioxide, the bubbly culprit behind fizzy drinks and baking magic. Our trusty sidekick here is limewater, a clear solution that turns milky white in the presence of CO_2. It's like watching a crystal-clear pond suddenly turn cloudy, signaling the arrival of our gaseous suspect.

Real-world applications: This test helps us identify carbonate minerals in rocks, monitor the bubbly action in bread dough, and even check if fire extinguishers are ready for action.

(c) Chlorine (Cl_2): The Vanishing Act

Chlorine, a potent gas used to disinfect water, takes center stage now. This time, our litmus paper performs a disappearing act, turning white as chlorine bleaches it away. It's like watching a colorful piece of evidence fade into thin air, leaving behind a blank slate.

Real-world applications: We use this test to keep our swimming pools safe, detect chlorine leaks in factories, and even analyze mysterious substances at crime scenes.

(d) Hydrogen (H2): The Explosive Escapade

Hold on tight, because hydrogen, the lightest and most abundant element in the universe, is about to make a loud entrance! A simple flame held near hydrogen ignites it with a satisfying "pop," like a tiny firecracker announcing its presence.

Real-world applications: This test helps us find hydrogen leaks in fuel cells and pipelines, ensuring safety and efficiency in these cutting-edge technologies.

(e) Oxygen (O2): The Rekindled Flame

Oxygen, the life-giving gas we breathe, is our next target. A glowing ember, on the verge of fading away, bursts back to life when it encounters oxygen, like a phoenix rising from the ashes.

Real-world applications: This test is crucial in hospitals to monitor oxygen levels for patients, in factories to ensure efficient combustion, and in environmental studies to assess the health of our planet.

(f) Sulfur Dioxide (SO2): The Color Thief

Our final case involves sulfur dioxide, a pungent gas released from volcanoes and industrial processes. This time, we'll use a vibrant purple solution of potassium manganate (VII) that loses its color in the presence of SO2, like a thief stealing away its vibrant hue.

Real-world applications: This test helps us track sulfur dioxide emissions from factories and volcanoes, ensuring clean air and a healthy environment.

Conclusion: The Power of Observation and Deduction

Just like a detective piece together clues to solve a case, we've used our chemical tests to identify these elusive gases. By observing color changes, precipitates, and fiery reactions, we've unlocked the secrets hidden within the air around us.

Remember, safety is paramount in our detective work. Always handle chemicals with care, wear protective gear, and perform experiments in well-ventilated areas.

Now go forth, budding gas detectives, and use your newfound knowledge to explore the fascinating world of chemistry!

Behold! The Alchemist's Fire: A Tale of Colorful Cations

Forget the sterile lab coats and safety goggles for a moment, and let's gather around the crackling fire like alchemists of old. Tonight, we're not brewing potions, but unlocking the secrets hidden within the very flames themselves. For within their incandescent dance lies a hidden language – a symphony of color that whispers the names of metallic elements.

The Dance of Electrons: A Fiery Ballet

Imagine tiny atoms, each with its own entourage of electrons whirling around like planets in a miniature solar system. When we introduce these atoms to the fiery embrace of a flame, something magical happens. The heat infuses the electrons with an almost ecstatic energy, causing them to leap to higher orbits, like dancers taking a dramatic leap across the stage.

But this excited state is fleeting. Like a graceful pirouette returning to earth, the electrons soon descend back to their original orbits, releasing their pent-up energy as bursts of light. And here's the secret: each element's electrons perform a unique dance, emitting light of a specific wavelength – a distinct color that reveals its identity.

A Kaleidoscope of Clues: Deciphering the Flame's Whisper

Let's peer into the flames and see what secrets they reveal:

Lithium: A passionate crimson red, like the heart of a ruby, bursts forth – the signature of lithium, a metal used in everything from batteries to mood stabilizers.
Sodium: An intense yellow, reminiscent of the midday sun, dominates the flame. Sodium, the ubiquitous element of salt and seawater, announces its presence with undeniable boldness.
Potassium: A delicate lilac, like the soft hue of twilight, emerges. Potassium, essential for life and found in bananas and potatoes, whispers its presence with a subtle beauty.
Calcium: A warm brick red, like the glow of a hearth fire, appears. Calcium, the building block of bones and teeth, reveals its strength in a steady, comforting hue.
Barium: A refreshing apple-green, like the first leaves of spring, flickers into view. Barium, used in medical imaging and fireworks, adds a touch of vibrant life to the flames.
Copper: A mesmerizing blue-green, like the depths of the ocean, shimmers before our eyes. Copper, the ancient metal of coins and wires, shows its versatility in a captivating blend of colors.

Beyond the Rainbow: The Flame's Limitations and Legacy

While the flame test is a powerful tool, it has its limits. Not all elements grace us with a colorful display, and sometimes, the fiery hues can mingle and mislead. But fear not, for this ancient technique has evolved. Modern instruments like flame photometers and atomic emission spectrometers amplify the flame's whispers, allowing us to measure and analyze the emitted light with incredible precision.

From classrooms to crime labs, the flame test continues to illuminate our understanding of the elements. It's a testament to the enduring power of observation and the beauty that lies hidden within the simplest of things – a dancing flame, a whisper of color, and the secrets of the elements revealed.

About Author

I am bestselling author. Data scientist. I have proven technical skills (MBA, ACCA (Knowledge Level), BBA, several Google certifications) to deliver insightful books with ten years of business experience. I have written and published 400 books as per Goodreads record.

ORCID: https://orcid.org/0009-0004-8629-830X

Azhar.sario@hotmail.co.uk

www.ingramcontent.com/pod-product-compliance
Lightning Source LLC
Chambersburg PA
CBHW031340230926
51848CB00002B/321
* 9 7 8 3 3 8 4 5 7 4 1 6 9 *